EARTH CYCLES

Recent Titles in
Greenwood Guides to Great Ideas in Science
Brian Baigrie, Series Editor

EARTH CYCLES

A Historical Perspective

David Oldroyd

Greenwood Guides to Great Ideas in Science
Brian Baigrie, Series Editor

GREENWOOD PRESS
Westport, Connecticut • London

Library of Congress Cataloging-in-Publication Data

Oldroyd, D. R. (David Roger)
   Earth cycles : a historical perspective / David Oldroyd.
        p.   cm.—(Greenwood guides to great ideas in science, ISSN: 1559–5374)
   Includes bibliographical references and index.
   ISBN 0–313–33229–0 (alk. paper)
   1. Geology—Periodicity.   2. Geological time.   3. Cycles.   4. Geology—Philosophy.
I. Title.   II. Series.
   QE33.2.P47O63   2006
   551—dc22   2006004787

British Library Cataloguing in Publication Data is available.

Library of Congress Catalog Card Number: 2006004787
ISBN: 0–313–33229–0
ISSN: 1559–5374

First published in 2006

Greenwood Press, 88 Post Road West, Westport, CT 06881
An imprint of Greenwood Publishing Group, Inc.
www.greenwood.com

Printed in the United States of America

The paper used in this book complies with the
Permanent Paper Standard issued by the National
Information Standards Organization (Z39.48–1984).

10 9 8 7 6 5 4 3 2 1

Every reasonable effort has been made to trace the owners of copyright materials in this book,
but in some instances this has proven impossible. The author(s) [editor(s)] and publisher will
be glad to receive information leading to a more complete acknowledgments in subsequent
printings of the book and in the meantime extend their apologies for any omissions.

CONTENTS

SERIES FOREWORD

The volumes in this series are devoted to concepts that are fundamental to different branches of the natural sciences—the gene, the quantum, geological cycles, planetary motion, evolution, the cosmos, and forces in nature, to name just a few. Although these volumes focus on the historical development of scientific ideas, the underlying hope of this series is that the reader will gain a deeper understanding of the process and spirit of scientific practice. In particular, in an age in which students and the public have been caught up in debates about controversial scientific ideas, it is hoped that readers of these volumes will better appreciate the provisional character of scientific truths by discovering the manner in which these truths were established.

The history of science as a distinctive field of inquiry can be traced to the early seventeenth century when scientists began to compose histories of their own fields. As early as 1601, the astronomer and mathematician, Johannes Kepler, composed a rich account of the use of hypotheses in astronomy. During the ensuing three centuries, these histories were increasingly integrated into elementary textbooks, the chief purpose of which was to pinpoint the dates of discoveries as a way of stamping out all too frequent propriety disputes, and to highlight the errors of predecessors and contemporaries. Indeed, historical introductions in scientific textbooks continued to be common well into the twentieth century. Scientists also increasingly wrote histories of their disciplines—separate from those that appeared in textbooks—to explain to a broad popular audience the basic concepts of their science.

The history of science remained under the auspices of scientists until the establishment of the field as a distinct professional activity in the middle of the twentieth century. As academic historians assumed control of history of science writing, they expended enormous energies in the attempt to forge a distinct and autonomous discipline. The result of this struggle to position the history of science as an intellectual endeavor that was valuable in its own right, and not merely in consequence of its ties to science, was that historical studies of the natural sciences were no longer composed with an eye toward educating a wide audience that included non-scientists, but instead were composed with the aim of being consumed by other professional historians of science. And as historical breadth was sacrificed for technical detail, the literature became increasingly daunting in its technical detail. While this scholarly work increased our understanding of the nature of science, the technical demands imposed on the reader had the unfortunate consequence of leaving behind the general reader.

As Series Editor, my ambition for these volumes is that they will combine the best of these two types of writing about the history of science. In step with the general introductions that we associate with historical writing by scientists, the purpose of these volumes is educational—they have been authored with the aim of making these concepts accessible to students—high school, college, and university—and to the general public. However, the scholars who have written these volumes are not only able to impart genuine enthusiasm for the science discussed in the volumes of this series, they can use the research and analytic skills that are the staples of any professional historian and philosopher of science to trace the development of these fundamental concepts. My hope is that a reader of these volumes will share some of the excitement of these scholars—for both science, and its history.

Brian Baigrie
University of Toronto
Series Editor

ACKNOWLEDGMENTS

I should like to express my sincere and grateful thanks to the following persons who have aided me in the preparation of this book: Brian Baigrie, David Branagan, Charles Drake, Ellen Drake, Kevin Downing, Philip Heckel, Linda Hinnov, Ted Lilley, Kerry Magruder, Andrew Miall, Christer Nordlund, Jane Oldroyd, Paul Pearson, John Pigott, and Tom Stanley. I benefited from a period of study at the excellent library of the University of Oklahoma, and the University of New South Wales continues to furnish me with all the necessary facilities to enable me to continue writing during my retirement.

ILLUSTRATION LIST

1

INTRODUCTION: THINKING IN CYCLES

There is an old joke about a story that can extend to infinity:

> It was a dark and stormy night, and the captain said to the mate: "Mate, tell us a story." And the story began thus:
> It was a dark and stormy night, and the captain said to the mate: "Mate, tell us a story." And the story began thus:
> . . . !

The joke of such a tale, with its infinitely long tail, quickly wears thin, however. As a story, it is pointless as a way of imparting useful information. It explains nothing; says almost nothing. It would be surprising if it had any sort of analogue in scientific explanation.

A tale can get more interesting and funnier if it takes longer to circle back on itself, as in the old Yorkshire song: "On Ilkley Moor bah't [without] 'at [hat]"

> Wheer hast tha' bin sin I saw thee? (I saw thee?)
> On Ilkley Moor bah't 'at
> Wheer has tha' bin sin I saw thee?
> Ditto
> On Ilkley Moor bah't 'at
> Ditto
> Ditto
> Ah've been a-courtin' Mary Jane (Mary Jane)
> On Ilkley Moor bah't 'at
> Ah've been a-courtin' Mary Jane
> Ditto
> On Ilkley Moor bah't 'at
> Ditto
> Ditto

Tha'll go an catch thee death o' cowld
On Ilkley Moor bah't 'at
Tha'll go an catch thee death o' cowld
(etc. . . .)
Then we shall 'ave to bury thee . . .
Then worms'll come and eat thee up . . .
Then ducks'll come and eat up t'worms . . .
Then we shall come and eat up t'ducks . . .
Then we shall all have etten thee . . .

This tale is different. It's a story with a beginning (going onto Ilkley Moor without adequate clothing, courting a young lady), a complicated and implausible causal sequence in the middle verses, and an unhappy ending (a young fellow being eaten by his relatives). To understand the history, the listener needs additional information such as the dangers of cold, burial practices, the eating habits of ducks and worms, and so on. In addition, there's a false assumption that eating some molecules from someone's decayed body is the same as eating him. The repetition of the refrain may suggest that the song is structurally analogous to the captain and mate story. But it isn't.

There's also the nursery rhyme "The House That Jack Built":

This is the malt
That lay in the house that Jack built.
This is the rat,
That ate the malt,
That lay in the house that Jack built.
This is the cat,
That chased the rat, that ate the malt,
That lay in the house that Jack built.
This is the dog that worried the cat,
That chased the rat, that ate the malt,
That lay in the house that Jack built.
This is the cow with the crumpled horn,
That tossed the dog, that worried the cat,
That chased the rat, that ate the malt,
That lay in the house that Jack built.

And so on, until:

This is the farmer sowing the corn,
That kept the cock that crowed in the morn,
That waked the priest all shaven and shorn,
That married the man all tattered and torn,
That kissed the maiden all forlorn,
That milked the cow with the crumpled horn,
That tossed the dog, that worried the cat,
That chased the rat, that ate the malt,
That lay in the house that Jack built.

This story builds up information but doesn't yield a *causal* sequence or a history. It describes the relationships of things in the farmyard. As told, there are nine "degrees of separation" between the farmer and Jack's house. There could be many other relationships. The farmer might simply be Jack's father.

Geology is a historical or "paletiological" science (it deals with "ancient causes"). So, of the three examples, it would seem to have closest analogy to the Yorkshire song. This is true, but it is only part of the story as far as geoscience is concerned.

In fact, some scientific theories are closer in structure to the captain's story than to the unhappy history of the Yorkshire swain or the way things were interrelated in Jack's farmyard. An example would be William Harvey's (1578–1657) theory of the circulation of the blood, published in his *De motu cordis* (1628). Blood is pumped from the heart through the arteries, but it circles back again through the capillaries and veins and does so continuously—as long as a person lives. Because of the circulation, there is no need to invoke any magical formation or disappearance of blood. Cyclicity gives the model its aesthetic appeal. The hydrological cycle, as proposed in antiquity by Aristotle, is analogous to Harvey's theory. Such cyclic theories are attractive because they need no beginning or end.

Isaac Newton's (1642–1727) theory of the workings of the solar system—developed in the latter half of the seventeenth century and based on the principles of his three laws of motion and gravitational "action at a distance"—envisaged a perpetual motion "machine" in which the planets and their satellites circled the Sun forever, in regular and law-like patterns that could be described mathematically. Predictions could be made years in advance about the seasons, eclipses, or positions of the planets and the stars in the sky. But Newton could not explain the beginning of it all in scientific terms. For this, he had to hypothesize a god that created matter, gave it its initial motion, and decreed the laws that it had to "obey." This "defect" or departure from cyclicity did not matter in the seventeenth century. People did not want God to be left out of the picture, and an astronomy that made possible explanation and precise prediction and that was largely intelligible (except for the "cause" of action at a distance) was highly satisfactory. Newtonian astronomy worked, without needing to know about the origin of mass, space, and time. Moreover, while there are many cyclic models or theories in science, like that of Harvey, all scientific theories, including geological theories, don't involve cycles or resemble the captain's story.

Newtonian time "flowed" in one direction, like a river at a constant speed. Nevertheless, cycles were involved in his cosmology: those of the day, the month, the seasons, and the year as well as what the ancients called the "Great Year," during which period (estimated in the ancient world to be about 36,000 years but now determined to be approximately 26,000 years) the position of the Sun in the sky at the time of the equinoxes slowly shifted against the background of the fixed stars before returning to its initial position (see Figure 9.2). Newton understood that such a phenomenon could occur because of the "wobble" of the Earth's axis in space (like that of a spinning top), giving rise to the "precession of the equinoxes."

In the ancient world, long before such things were understood à la Newton, the Great Year was known through the records of astronomical observations and was associated with thoughts about the cyclic recurrence of world events. Cycles were, in fact, to be expected. There was the cycle of the seasons of the year, of work tasks, of life and death, or of the seven ages of man, as described by Shakespeare (or Monty Python). It seemed natural and right, from an anthropomorphic perspective, that changes in the cosmos should be cyclic. This could be—and was—given mystical or mythopoeic significance.

Thus, Mircea Eliade (1959), a notable French writer on myths, distinguished between "sacred" and "profane" time. "Circular" time was expressed in religious festivals, such as the New Year, which celebrated the origin of the cosmos (or Earth), with an annual renewal. The cosmos was supposedly recreated each year, at which time it regained its original sacred state. (A faint shadow of this survives in New Year resolutions.) The rituals of the festivals of "sacred time" supposedly allowed a restoration of order rather than chaos, the behavior of the gods being emulated in the festival rites. "Profane" time prevailed during the rest of the year, as the time of everyday life. Thus, in a sense, archaic man lived in a continual present (like an animal?), for the past prefigured the future, and no transformation was irreversible. The idea of repeated cosmic birth, death, and rebirth was presumably prompted by the phases of the Moon and the repetition of seasons. Such ideas, Eliade showed, occur in cultures around the world.

The "model" could also take the form of thinking not only that the cosmos was sacred and that mankind could be rejuvenated by annual rites but also that time *itself* was cyclic (as in the Great Year). So we have the notion of the inexorable wheel of fate. This is found in Greek tragedy, when a person struggles with but is unable to escape his or her fate or destiny. Such views have been expressed in various literary ways in, for example, the Oedipus myth or Shakespeare's *Julius Caesar* (as in the preordained fate of the "Ides of March").

Such ideas are barely comprehensible to moderns. Taken to their extreme, they might imply that I would be typing this very sentence 36,000 years from now. If taken literally and if this sentence were to reemerge endlessly every 36,000 years, it would seem a dismal prospect. I could not have the pleasures (and agonies) of free will. This paradox would seem to underlie the anguish of Greek tragedy (and much subsequent literature, too), though probably few have seriously believed that cycles repeat themselves in *every* detail.

Historian of science Stanley Jaki has suggested that the replacement of the idea of cyclic time was a precondition for the emergence of science and the modern world. He discussed what he called the "Treadmill of the Yugas," describing the cosmology and ideas of time in ancient India (which survives in the modern world in the thinking of, for example, the Hare Krishna movement). The Hindu beliefs were profligate with time. As summarized by Jaki (1974, 3), there were the following cycles:

1 divine day	1 [astronomical] year = 360 days
1 divine year	360 years
12,000 divine years	4,320,000 years
= a cycle of four *yugas*	
= 1 *mahayuga*	
1,000 *mahayugas*	4.32×10^9 years
= 1 *kalpa*	
2 *kalpas*	8.64×10^9 years
1 day of Brahma	3.1104×10^{12} years
= 1 year of Brahma	
= 360 days of Brahma	
100 years of Brahma	3.1104×10^{14} years
= 1 life of Brahma	= about 311 trillion years

These large, almost incomprehensible numbers were evidently "drawn from a hat" and had no scientific significance.

Brahma (the Creator of the universe) was one of the three main deities in the Hindu pantheon, the other two being Vishnu (the Preserver) and Shiva (the Destroyer). At the completion of the four *yugas*, Shiva supposedly destroys the universe in the "night of Brahma." But Brahma then re-creates it in a new "day." And so the cycles continue. Even at the end of the "life of Brahma," the process simply starts again. There was supposedly a successive degeneration of society and moral character during the four *yugas*, regarded metaphorically as the Golden, Silver, Bronze, and Iron Ages. Again, the idea of the birth, growth, decline, and destruction of society, followed by a rebirth, may have been prompted by knowledge of the phases of the Moon.

Such prescientific ideas seem absurd when baldly presented thus. But they reveal that, in ancient India, beliefs about the magnitude of time were not dissimilar to modern ideas about the age of the universe. And perhaps—if "big-bang" cosmology is one day superseded by a cyclic theory of the expansion and contraction of the universe—modern science may then line up with early Indian metaphysical speculation regarding time and cycles. For we may find the idea of creation ex nihilo so obnoxious that a cyclic cosmology may be sought and eventually come to form the scientific consensus. If that happens, however, the similarity would be coincidental, for a cyclic cosmology must comply with modern theory and observational evidence as well as metaphysical preferences!

I make no judgment on this matter except to say that I find the creation of a cosmos out of nothing incomprehensible and that to explain such a mystery by invoking another one (God) gets one nowhere philosophically. But an infinitely old cosmos makes no sense to me, either. Be that as it may, the ancient Hindus had ideas on time that were more in line with modern geology than was the young-Earth time scale of the Judeo-Christian tradition, which came to prevail in the Western world and survives to the present in some quarters, particularly in the United States. Thus, when Charles Lyell (see Chapter 6) wrote his *Principles of Geology* (1830–1833), which argued for a vast geological time scale, with some cyclic features, it is significant that his introduction began by discussing Hindu ideas about time and cycles of destruction and renovation, presumably with the implication that an immense age for the Earth was an idea with a long and distinguished ancestry.

By contrast with Hindu theology, in Judaism a single god supposedly manifested "himself" in a "single" historical time, which, however, had a distinct beginning and end. For Christian theology, all time and history were sacred in that history was linear: from a special creation of the cosmos and the Earth, the settlement of mankind and other living creatures on the Earth, the coming of Christ, a second coming, and a state of bliss for those who "make it" to heaven at the Day of Judgment—which would be the end of time as we know it. In this "linear" history, there is no wheel of fate. We can, by our exertions, achieve or fail to achieve our goals. This metaphysic can—and eventually did—lead to the "idea of progress" in the eighteenth-century Enlightenment and the modern world. Jaki saw "linear time" as fundamentally "energizing." This may be so. But we need not dwell further on such metaphysical or theological matters.

Part of the process of achieving "progress" (in "linear" time) involved the exploration of our world, both in space and in time. The latter takes us into the world of geology. Evidently, some parts of geology, such as cycles of glaciation and amelioration of climate, sedimentary cycles, or cycles of mountain building, are essentially concerned with cyclic processes. Other parts, such as paleontology and the evolutionary histories of organisms, are profoundly noncyclic. Thus, according to a law proposed by the Belgian dinosaur hunter Louis Dollo (1857–1931) in 1890 ("Dollo's law"), Darwinian evolution never reproduces the same species, no matter that almost identical environmental conditions may recur.

I do not know the proportion of geoscience that is concerned with cyclic processes, but whatever it may be, cyclic theories are *essential* to modern geological theorizing and explanation. Moreover, they have the aesthetic appeal that Harvey saw in his theory. The emergence of geology as a science had to do, in part, with a "contest" between linear and cyclic theories. That contest has now run its course, and the two "themata" (see Chapter 14) now coexist comfortably. Geological cycles are not—needless to say—of a nature such that there will, at some time far into the future, be an identical river valley as that in which my house now sits as I type this sentence. Moreover, I have no fear of geological cycles and do not struggle against them as did the Greek heroes against their fates. Even so, cycles should perhaps cause us concern. What will happen to civilization if there is a renewed Ice Age? Or runaway global warming?

Leaving that aside, this book is concerned to outline the history of cyclic theories as they have appeared in and contributed to geoscience. A turn to thinking in millions or billions—rather than thousands—of years for the age of the Earth and thinking of cyclic processes occurring within the uniform flow of "Newtonian" time was an essential step for geological understanding, no matter that it produced something that was comparable in time scale to early Hindu thinking. What was a huge imaginative leap for those steeped in the Judeo-Christian tradition would, I dare say, have seemed commonsensical to the ancient Hindus. But ideas can recur in cycles, as do physical processes.

In the present book, we shall encounter patterns of relationships analogous to those in the story of Jack and his house (e.g., Rock A has been deposited on Rock B, which in turn has been deposited on Rock C and so on), ones like the "bah't 'at" song (e.g., *this* plate movement occurred and *caused* the formation of *that* mountain range, but continental rearrangements can eventually restore earlier configurations), and cycles of events like the captain and the mate story (e.g., the "geostrophic cycle"; see Figure 5.1). The first two kinds of pattern may occur within overarching cyclic geological process, and it is such cyclic concepts that we focus on here. But the cycles are never quite "perfect." They always have a "spiral twist." So the Earth *does* have a unique history, and geology can be—and is— a historical enterprise. I shall not write this sentence again in 36,000 years or after 311 trillion years!

2

IDEAS OF GEOCYCLES FROM THE ANCIENT WORLD TO THE RENAISSANCE

While important arts and crafts were developed in the East and there were significant developments such as the "invention" of zero, most historians of science regard the ancient Greeks as being the first to produce hypotheses about the physical world that were subjected to rational criticism or empirical test. So the Greeks are usually credited with originating modern science and philosophy—though the two were not then regarded as distinct. Thus, a number of ideas were proposed about changes of sea level or interchanges of land and sea, some based on geographical or historical information and others proposed in the course of "philosophical" discussion. Indeed, various hypotheses that might be called "theories of the Earth" were propounded.

The founder of the "Milesian school" is said to have been Thales of Miletus (ca. 624–548/547 B.C.), Miletus being located in what is now western Turkey. Thales knew of the approximate 19-year cycle of lunar eclipses, and (perhaps by luck rather than judgment) reputedly predicted the year of a solar eclipse. He is particularly remembered for suggesting that water was the fundamental "stuff" of the world. The idea that all things are, at bottom, one kind of substance was a remarkable hypothesis. Further, Thales suggested that the Earth floated on an immense ocean, with earthquakes being due to occasional "wobbles" of this great "island." This suggestion was, however, queried by his disciple Anaximander (ca. 611–547 B.C.), who asked, What held up the water? and, further, Of what is water itself made? Anaximander is also believed to have discovered the equinoxes and solstices (though they were probably known to the people who built structures like Stonehenge). He used and perhaps invented a sundial and produced a map of the known world and a geographical work, *On Nature,* that raised the issue of changing levels of land and sea (Heidel 1921). It was thought that the island of Rhodes (off the southwestern coast of Turkey) had emerged from the sea and that this might have been due to the fall of the sea level, the rise of land, or a combination of the two. Anaximander also knew of the growth of the Nile Delta. According to the geographer Strabo (64/63 B.C.–A.D. 24) (1960, 181), the fifth-century B.C. historian Xanthus of Lydia reported that stones in the form of marine shells (and impressions thereof) had been found in Armenia, Matiene, and Phrygia, far from the sea.

Anaximander's student Xenophenes of Colophon (ca. 540–537 B.C.) suggested that there were two fundamental "principles": earth and water. He was interested in meteorological phenomena and suggested that the Sun's heat drew moisture from the sea, which caused cloud formation and winds, and that water returned to Earth as rain. Thus, he had the idea of the hydrological cycle while fancifully suggesting that the Moon consisted of condensed moisture. Further, he envisaged repeated transgressions and regressions of the sea, leading to either a universal deluge or complete desiccation. There were thus cycles, but while he was close to a "Harveyan" cycle in his meteorology, the repeated desiccations and inundations were more like "swings of a pendulum."

No original writings survive from the mathematician Pythagoras of Samos (ca. 560–ca. 480B.C.), who taught a semimystical philosophy to the "Pythagorean school." Centered at Croton in southern Italy, the Pythagoreans believed that the ultimate reality of things "existed" in the form of numbers. (This has a parallel in modern science, where the ultimate "reality" of subatomic particles has sometimes been said to consist of their mathematical formalisms.) The Pythagoreans believed in a geocentric cosmos, with the Earth having a central fire. This idea has, in a sense, survived through to the present, though the historical chain of relationships is neither complete nor single, as in the case of "The House That Jack Built." Nor were Pythagorean ideas typically cyclic as far as their thinking about the Earth was concerned—though they thought that civilizations rise and fall in a cyclical fashion, typically being terminated by natural disasters.

The Pythagorean Empedocles of Agrigentum in Sicily (ca. 490–ca. 430B.C.) took a keen interest in the Earth's interior—and is said to have died in the crater of Etna! His ideas mingled religious, moral, and naturalistic aspects. Supposedly, there were two fundamental forces in nature ("Love" and "Strife"), operating in a universe of four elements, or "roots of things" (earth, water, air, and fire)—an idea that persisted variously until the late eighteenth century. Empedocles' elements seem to have been particulate, with pores, which allowed the formation of "compounds" when the particles were of the appropriate dimensions or shapes. Things also gave off "effluences," so that there was perpetual flux and movement, leading to destruction and decay; but this process could be reversed by the role of "Love," when objects would draw matter in (Guthrie 1965, 151). Like Anaximander, Empedocles thought the Earth was at the center of the cosmos not because it was supported but because there was no reason why it should move one way or another, being at the center of the "cosmic whorl." The sea, he thought, was "sweated" out of the Earth and was salty, like animal sweat. Unsurprisingly, given that Empedocles lived in Sicily, he thought that fires burned in cavities within the Earth. Indeed, there could be internal rivers of fire that sometimes issued from volcanoes. The internal fires could also heat subterranean waters.

Herodotus (ca. 484–ca. 425 B.C.) was a historian who detailed the wars of the ancient world; but in the second book of his *Histories,* he wrote on the physical and cultural geography of Egypt and suggested that the Nile Valley and the gulf of the Red Sea were analogous. They were so close at their north-ern ends that if the Nile's waters were diverted into the Red Sea, the latter waterway might become silted up in 20,000 years or less. So Egypt could have been produced by deposition of silt in the Nile Valley. The present extension of the Nile Delta northward was evidenced by sediment dredged up offshore. Herodotus also observed marine shells on the Egyptian hills and noticed that salt exuded from the soil to such an extent that it was damaging the Pyramids. The rich black soil of the Nile Valley was alluvial. Evidently, he envisaged long, slow "geological" processes that might alter the relative bounds of land and sea. But cycles were not proposed.

The most important philosophers of ancient Greece were, of course, Plato (428/427–348/347 B.C.) and his "student" Aristotle (384–322 B.C.), whose interests were more concerned with the natural world than were those of Plato. Nevertheless, in his dialogue *Phaedo* (109–13), Plato offered an account of the Earth's interior. He imagined its appearance from outer space (as we might say): it would be "like one of those balls which are made of twelve pieces of leather" (a dodecahedron, approximating a globe?), with various surface hol-lows. From these, passages supposedly led into the Earth through which water flowed back and forth. At the center was the chamber of Tartarus (as described by Homer). Air also flowed back and forth in the channels, driven along by the waters; and there were "great rivers of fire and streams of liquid mud, thin or thick (like the rivers of mud in Sicily, and the lava streams that follow them)." The inward and outward movements were likened to breathing. Moreover,

> [W]hen the waters retire from the regions below into the streams on the further side of the Earth, and fill them up like water raised by a pump, and then when they leave these regions and rush back hither, they again fill the streams here, and these being filled flow through subterranean channels and find their way to the appointed places, forming seas, and lakes, and rivers, and springs. Thence they again enter the Earth, some of them making a long circuit into many lands, others going to a few places and not so distant. (*Phaedo,* 112c)

For Plato, the underworld was the place where dead souls ended up before their reincarnation or permanent incarceration. But, for our purposes, it is the cyclic theory that is interesting. Like Harvey, Plato seemingly recognized the conceptual virtues of a cyclic physical theory. But the exact nature of the cycle has been debated. Şengör (2003, 41) has argued that in Plato's thinking there was only one major channel between the innermost depth, Tartarus, and the outer surface. Tartarus was a kind of "cul-de-sac." If this were not so, Şengör argues, the breathing analogy would not be "valid." But the previously cited quote seems to suggest a passage for transport of water from one side of the Earth to the other, which has been the interpretation of earlier scholars. Be this as it may, Şengör rightly points out that the limestone areas around the eastern Mediterranean, forming so-called karst(ic) topography with underground rivers, made Plato's "theory" physically plausible. Şengör also suggests the influence of Empedocles on Plato's model.

Aristotle regarded the cosmos as eternal, with the Earth at the center of the sphere of fixed stars. The infinitely old cosmos was held together—its opposed elements in harmonious balance with one another—by a nonpersonal god, envisaged as an "unmoved mover," which (*not* who) maintained the motion of the whole. Aristotle believed that each element had its natural place in the cosmos (fire above air, air above water, and water above earth). So, given an infinitely old Earth, one might expect the elements to have settled out thus. But things sometimes have "nonnatural" motions, being disturbed from their natural places, and then return thereto in an almost infinite number of subsidiary cycles. Besides, it was obvious that there was growth and decay on the Earth. So, as Aristotle put it in *Coming-to-be and Passing Away* (338a),

> If, then, the coming-to-be of anything is absolutely necessary, it must be cyclical and return upon itself; for coming-to-be must either have a limit or not have a limit, and if it has a limit, it must proceed either in a straight line or in a circle. But of these alternatives, if it is to be eternal, it cannot proceed in a straight line, because it can have no source, whether we take the numbers of the series downwards as future events or upwards as past events. But there must be a source of coming-to-be, though without coming-to-be itself being limited, and it must be eternal. Therefore, it must be a cyclical process. (Aristotle 1965, 325, 327)

In this logic-chopping way, Aristotle argued that processes in the cosmos *had* to be cyclic. "Absolute necessity is present" by courtesy of cyclical processes. Such reasoning could have appealed to Harvey (and very likely actually did so).

It comes as no surprise, then, that we find cyclical processes in Aristotle's astronomy and in his ideas on meteorological phenomena. He described the hydrological cycle, though he thought it was driven by the annual cycle of the seasons rather than daily weather activity (*Meteorologica,* bk. 1, chap. 9). (But such a model was suited to the Mediterranean climate with hot, dry summers and cooler, rainy winters.)

It has also been claimed by Şengör (2003) that the *Meteorologica* contained "the first theory of global tectonics" (earth movements). Aristotle accepted that the Earth had internal cavities and passages, thus accounting for the fact that the Aral and Caspian seas, for example, had rivers entering but not exiting. He also thought that winds could blow in the internal passages, causing earthquakes. Additionally, Aristotle accepted the kind of evidence given by Herodotus (without specifically mentioning him) to the effect that dry land could be inundated or seas in-filled by sediments, at least locally. Such changes were so slow that they got forgotten over many generations. But they occurred in an "orderly cycle" (*Meteorologica*, 351a). The deposition of sediment at place *A* would cause a marine transgression at some other locality (*B*). But if rivers dried up at *A*, then the sediment there deposited might be inundated by the sea-level rise, caused by sedimentation elsewhere (*B* or *C*?). If pursued on a large scale over a great length of time, then the locations of the seas might shift. Such changes might be driven by climatic changes occurring in different parts of the world. Aristotle seemed to imply (*Meteorologica*, 366a) that earthquakes would be most severe near coasts, as the sea might drive escaping wind back into the Earth, causing greater stresses. In addition, earthquakes would be severe at localities where the Earth is particularly porous.

Aristotle distinguished between the Earth's *dry* and *moist* exhalations. The latter were involved in the formation of rain in the hydrological cycle, while the former were supposedly involved in the Earth's internal winds. These caused earthquakes, which were more severe where the sea currents were strongest and where the Earth below was hollow or porous. It was claimed that winds burst out of the Earth when volcanic eruptions occurred, and the "percussion" (friction?) generated heat. If a wind causing an earthquake is opposed by one blowing over the sea, then a tidal wave (tsunami?) could be produced. Earthquakes were also supposedly linked to periods of heavy rain or drought (*Meteorologica*, 366b). If the Earth's pores are filled with water, there will necessarily be increased exhalation. Winds were strongest during droughts.

On the basis of such considerations, Şengör (2003, 44–45) has proposed a cyclic theory of the Earth for Aristotle. A region subjected to heavy rain has its pores filled with water, and the land is elevated. An adjacent basin can develop with closed pores due to former drought and land collapse. This basin can become occupied by the sea as heavy rain begins to fall in that area, which will then begin to swell and become elevated by filling of the pores. By then, the former high land will be subsiding through loss of water from its pores, exhalation, evaporation, and runoff. Thus there will be a kind of seesaw effect, raising and lowering two adjacent areas. Earthquakes and volcanoes are concentrated in the region between the two.

As far as I know, Şengör's is the first attempt to interpret the *Meteorologica* as proposing an unending sequence of tectonic movements, with interchanges of land and sea. It is possible that Aristotle had some such theory in mind, but we are, I think, offered a rather imaginative reading of his text. Aristotle undoubtedly envisaged cyclic interchanges of areas of land and sea, though I question whether the interchanges were integrated as closely with his theory of earthquakes and climate changes as Şengör suggests.

Aristotle's theories were subsequently developed by his pupil Theophrastus (ca. 372/369–ca. 288/285 B.C.), who may have written book 4 of Aristotle's *Meteorologica*. (But this had little "geological" interest.) Theophrastus is remembered chiefly for his botanical work and a work on minerals. Strato(n) of Lampsakos (ca. 335–269 B.C.), another Greek colony in Asia Minor, succeeded Theophrastus at Aristotle's Lyceum in Athens. His works are lost, but there are references to his ideas in later texts. For example, he is mentioned in Strabo's *Geography* (bk. 1, 3.4), where it is suggested that rivers flowing into the Black Sea had caused it to overflow, cutting the Bosphorus channel. Strato also envisaged a time when the Black Sea would be filled with sediment. In Egypt, there was good evidence for the retreat of the sea. Strato also thought that the ocean floors could be as uneven as land surfaces. He apparently gave an excellent account of the formation of a submarine delta and has been praised by the French geohistorian François Ellenberger (1996, 15) for producing "an amazing uniformitarian text" (for discussion of uniformitarianism, see Chapter 6).

The geographer Strabo was born in Pontus (Turkey). He resided chiefly in Alexandria but traveled widely, from the Black Sea to Ethiopia. He cited the occurrence of marine shells near the Temple of Ammon (the Egyptian "ram-god," depicted as having curled ram horns on a human head, resembling what were later known as ammonites), about 600 kilometers up the Nile Valley at Thebes (Strabo, bk. 1, 3.3; bk. 17, 3.11). The astronomer and geographer Eratosthenes (276–195 B.C.) had previously thought the shells suggested the withdrawal of the sea from the area. He had envisaged small-scale elevations or depressions of land and associated such changes with what we would call seismic activity. But such changes in Egypt were, Strabo thought, only temporary and not really cyclic. By contrast, in book 17 (1.36), he spoke of global changes, with plains becoming mountains and vice versa or springs and rivers forming or disappearing. Further, he suggested that such changes in the areas of land and sea might be due to the rise or fall of the land beneath the seas (bk. 1, 3.5; 3.10). He reported the appearance of new islands in the Mediterranean, associated with volcanic outbursts, and supposed that such things might have occurred extensively over time, indicating uplift from—or of—the seabed.

The celebrated Roman poet Publius Ovidius Naso, or "Ovid" (43 B.C.–A.D.17/18), who came from Sulmo(na), Italy, studied rhetoric and law and started in public administration but turned to writing as a career. He composed poetry and wrote at least one play, but his best-known work is his *Metamorphoses,* in which he assembled numerous myths and legends, with the common factor that they all involved transmutations. Ovid was not a naturalist, but he gave poetic expression to many common beliefs of his time. Book 1 of *Metamorphoses* gives an account of the divine Creation from Chaos, which is not vastly different from that of *Genesis* but was the product was an Aristotelian-style universe of immense age.

Book 15 concluded the work with some Pythagorean ideas:

> Nothing is constant in the whole world. Everything is in a state of flux, and comes into being as a transient appearance. Time itself flows on with constant motion, just like a river: for no more than a river can the fleeting hour stand still. (Ovid 1955, 369)

The four elements were constantly changing: earth to water, water to air, air to fire, and back again, in reverse order. All could change, but "the sum of things remain[ed] unchanged." Ovid claimed to know of land changing into sea and vice versa, referring to seashells found far from the ocean. Valleys could be excavated by running water. Mountains could be transferred to plains by floods. Rivers or springs could appear or disappear or change from fresh to saline. New hills had appeared in what had been plains (Ovid 1955, 371–72). This was not a cyclic geological theory per se, but it offered some "geomorphological" ideas that were not generally accepted in the West until perhaps the nineteenth century. We shall see further (p. 40) that later writers found stories in *Metamorphoses* that might be construed as accounts of significant geological events, presented in mythopoeic form.

The Roman politician, orator, playwright, and Stoic philosopher Lucius Seneca (4 B.C.–A.D.65), who had the misfortune to be called to serve as tutor to Nero, composed a treatise on natural history titled *Quaestiones naturales* (*Natural Questions* or *Physical Questions*) (Seneca 1971–1972), which covered much of the ground mentioned previously. After Nero became emperor, Seneca was the "power behind the throne" for several years. But later he was implicated in the murder of Nero's mother, following which the emperor abandoned himself to his well-known excesses and eventually determined to rid himself of Seneca. He was accused of plotting Nero's death and had to commit suicide by cutting his veins!

Some of Seneca's treatise had to do with moral and philosophical questions, but most sections dealt with phenomena such as discussed in Aristotle's *Meteorologica,* on which it was based to a considerable extent. Thus, like Aristotle, Seneca believed that the Earth contained passages and cavities and that the winds therein produced internal fires, earthquakes, and volcanoes. But he added the notion of "tension," mediated chiefly by the air, as an explanatory principle, with the idea that there were grand cycles, supposedly involving increases and decreases of cosmic "tension" (*tonos*).

Quaestiones, book 3, dealt with different terrestrial waters. Seneca believed that the four elements are interchangeable, which could explain why he didn't think rivers originated as rain. The water could just as easily have come from within the Earth or by transformation of the other elements. So, he thought, it was the fate of the Earth and living organisms to be overtaken by a great deluge, when all but the highest peaks would be inundated. This would occur in the natural course of events as the cosmic "tension" relaxed. All would become one inchoate watery mass, and humanity would be destroyed. But following this destruction, the Earth would absorb the waters and drive them back to their "secret dwelling-place." Then organisms would be renewed, and "men ignorant of sin" would be re-created.

There was also a grand cycle of the elements, occurring during a Great Year. During the first phase, tension is at its maximum so that only the "world-soul" (God) exists. Then, with cooling, the fiery matter passes into air, air into water, and water into earth. In the reverse phase, earth produces water, water air, and air fire. All this was not described in the *Natural Questions* (which did, however, give attention to earthquakes and volcanoes, in which Seneca had a special interest). It is a nice question which phase of the cycle we are in at present. Perhaps the "Noachian" flood of legend was the "aqueous phase"? If so, we are now on the way to some future conflagration. (There were other flood legends in the ancient world besides that of *Genesis*.)

Thus, Seneca's world process was circular. The changes were deemed to be inexorable: the whole constituted the great wheel of fate. But were that so, one may wonder why Seneca sought to involve himself in politics and, for a time, enrich himself. Perhaps he saw the inconsistency himself and for that reason withdrew from public life and become an ascetic. Be that as it may, we can see Seneca offering a more "philosophical" view of matters that Ovid treated as commonplace beliefs.

The Stoic picture was given greater clarity by the grammarian and philosopher Ambrosius Macrobius (ca. 430 A.D.) in his *Commentary on the Dream of Scipio*—a work that gave a generalized account of Stoic cosmology. He wrote,

> [O]r civilizations often perish almost completely, and they rise again when floods or conflagrations in their turn subside. . . . [P]hilosophers have taught us that ethereal fire feeds upon moisture, declaring that directly under the torrid zone of the celestial sphere . . . nature placed Ocean . . . in order that the whole broad belt over which the sun, moon, and the five errant planets travel might have nourishment of moisture from beneath. . . . Since heat is nourished by moisture, there is an alternation set up so that now heat, now moisture predominates. The result is that fire, amply fed, reaches huge proportions, and the moisture is drained up. The atmosphere in this state lends itself readily to conflagration, and the Earth far and wide is ablaze with raging flames; presently their progress is checked and the waters gradually recover their strength. . . . Then, after a great interval of time, the moisture thus increasing prevails . . . so that a flood covers the lands, and again fire gradually resumes its place. (Macrobius 1952, 218)

Macrobius's scheme thus described was clearer and simpler than Seneca's and was shorn of metaphysical or moral dimensions. It was certainly cyclic, but it was not attached to any observational basis (though it almost fit the Australian environment—the land of floods and droughts!). Macrobius stood in the tradition of speculative philosophy, not geoscience. For all their writing, arguing, and empirical knowledge in the ancient world, no one (as far as records reveal) developed a successful "theory of the Earth," either linear or cyclic. There are events described in the Old Testament that may plausibly be accounted for by natural "geoprocesses," such as the angel with the flaming sword (*Genesis* 3: 24), which was probably "a way of talking about" a gas flare in the Arabian desert. But mythology, religion, or philosophy dominated thinking about such matters. Geological phenomena were known and described, the world around the Mediterranean Sea was explored, and some maps were produced. But no geoscience proper has survived from early times. And probably none was produced.

Writers such as Seneca and Macrobius were concerned with the rise and fall of civilizations, and by Macrobius's time the Roman Empire was in decline. With its demise, the world of speculative and empirical learning contracted and was resurrected only by the rise of the Muslim civilization in the early Middle Ages. Not much is known to geohistorians about the geoscientific world from that civilization, but a multiauthored encyclopedic work from the tenth century, probably compiled in Basra and known as *Rasā'el Ikhwān al-Ṣafā' (Epistles to the Brethren of Purity)*, contained a cyclic theory that was in advance of anything from the Greeks or Romans.

The authors gave a Great Year as 36,000 years (3,000 years for each sign of the Zodiac), during which period there was a large geocycle. What we would call weathering and erosion occurred, with transport and deposition of sediment, layer by layer, in seas and lakes. The layers formed submarine hills, but the deposition caused water to overflow onto the land. Regions that were land became seas and vice versa. Then the new mountains would be worn down in a subsequent cycle (see Ellenberger 1996, 64–65).

This was almost a modern geocyclic theory, with an open-ended time frame, except that there was surely a mystery about the process(es) of uplift or mountain formation. But *this* mystery took long to solve, and the effort to provide a complete solution (or tectonic theory) to everyone's satisfaction continues, even to the present. It is not known whether these Arabian ideas were passed to the West. They may have reached Europe via Moorish Spain. Whether they moved further is not known.

But Aristotle's ideas moved into medieval Europe by translations of his works, in the first instance in Sicily and Spain, and sometimes by Arabic translations. Christianity was well established in Europe, and universities appeared, teaching clerics, lawyers, and doctors. By the thirteenth century, a considerable number of Greek texts had become available in Latin translation and were the subject of extensive "commentaries," rationalizations, or disquisitions as the basis of higher education. One of the great intellectual problems was to reconcile Aristotle's ideas and Christian doctrines—a difficult undertaking, given their entirely different ideas about time and the nature of God (an "unmoved mover"—or a Creator that took a personal interest in human affairs and behavior?). The intellectual contortions that the reconciliation involved need not detain us here. The Aristotelian geocentric cosmos was *not* a problem. On the contrary, it was congruent with a divinely created world, with "Man" at the central position.

However, in the work of the logician Jean Buridan (ca. 1297–ca. 1358), a professor in the Faculty of Arts at the University of Paris and rector of the university from 1327, we find sophisticated modifications of the Aristotelian cosmology that gave rise to a remarkable "theory of the Earth" with cyclic features. It was contained in Buridan's commentary on Aristotle's *Meteorologica*, and the relevant parts were published in French by the nineteenth-century historian and philosopher of science Pierre Duhem (1958, 293–305) (see also Shapiro 1964, 542–47). Buridan's ideas also appeared in a commentary on

Aristotle's *De caelo* (Buridan 1942), and parts are available in English (Grant 1974, 621–24).

Buridan addressed the old question of whether land and sea had always retained their respective positions. If such movements occurred, they could not, he pointed out, have been associated with celestial motions such as those associated with the Great Year. These were too rapid, and the putative changes on Earth would have been noticed during historical time. Moreover, the seas could not have overtopped the mountains during the deluge. (He estimated that the mountains of France were much higher than they actually are.)

It seemed to Buridan, though, that the Atlantic side of the globe was chiefly water, while the other side (Eurasia and Africa) was chiefly land. In addition, it was supposed that there were subterranean waters—an assumption that was essential to his theory. Further, Buridan knew that rivers carry materials from the land to the sea, so that the part of the Earth under the sea is constantly being increased by sedimentation, while the (supposedly opposite) land area is diminished. But, according to Aristotelian cosmology, the Earth's center of gravity must always be located at or tend to move toward the center of the universe (or the sphere of fixed stars). Consequently (by a kind of "balance" argument), as the "ocean" side of the Earth increases by sedimentation, the "land" side must be *elevated* so as to maintain the position of the center of gravity of the whole with respect to the celestial sphere. Elevation on the landward side would be further increased by the exposure of the land to the heat of the Sun, which might make it expand like heated dough, becoming "rarer and lighter."

In consequence, given enough time (of which Aristotle allowed an infinity), the parts of the Earth now under the sea would gradually move toward its present center, and the materials presently at the Earth's center would slowly come to the surface of the side of the Earth's landward side, whence they could be returned to the "ocean" side by weathering, erosion, and fluvial transportation. Thus, terrestrial matter would undergo a perpetual cycling process. But some parts of the "land" side are evidently harder than others, and the softer parts get worn away preferentially. Hence, mountains and valleys are formed by differential erosion on the "land" side of the Earth.

Further, Buridan envisaged a slow progression of the "ocean" side of the Earth around the globe, though it is not clear exactly why it would change thus. (I would have thought that it could equally well have taken a "random walk"!) Buridan added that since one could consider the direction of a meridional line at a particular locality to be related to its distance from the ocean, so one might entertain the notion of shifts in meridional directions relative to the unchanging celestial sphere. Another scholar at Paris University, Nicole Oresme (1323–1382), suggested that the transfer of sediment from one side of the globe to the other might lead to the movement of the poles on the surface of the Earth (Duhem 1958, 306), an idea that came to have a long history (see Chapter 4).

Obviously, such theories required enormous "drafts of time," rivaling the numbers of years for a Hindu cosmic cycle, and it was consequently incompatible with the Christian notion of time (which Buridan was supposed to accept). He said, in response, that there were the knowledges of reason and faith and that he relied on the latter while contemplating the former. This, of course, allowed him to speculate about physical matters as much as he might wish! Using this "debating trick," Buridan claimed that his "theory of the Earth" was simply a hypothesis that allowed "all the phenomena that are apparent to us . . . [to] be saved" (Shapiro 1964, 542). Buridan's separation of faith and reason seems to have kept him safe from theological criticism, though the distinction later proved insufficient to prevent Galileo in the seventeenth century from being severely rebuked for holding heliocentric views to the point of being shown some instruments of torture.

Buridan's remarkable theory was supported by his pupil Albert of Saxony (1316–1390) and was transmitted to Renaissance scholars through his writings (Albert of Saxony, 1492; Duhem 1958, 309–12), and Buridan's own text was eventually printed in 1516 (not seen). It is hard to determine what influence it may or may not have had on subsequent thinking. But Duhem (1958, 309–23) referred to a line of philosophers—Thimonis, John Major, William of Heytesbury, Thomas Judaeus, Marsilius of Inghen, Pierre d'Ailly, Leonardo da Vinci, sixteenth- and seventeenth-century Jesuits, and eventually Leibniz—who considered Buridan's theory of the Earth significant.

By way of example, I mention some ideas of Leonardo da Vinci (1452–1519). As is well known, many of his ideas and inventions were developed in private, being recorded in manuscripts in which the writing was in "mirror" form (he was left-handed, so by writing from right to left he avoided smudging his ink) and were for long undeciphered. However, they are now published, and one of them, titled "Codex Leicester" (written in Milan, 1506–1510, and presently owned by Bill Gates) (Leonardo 2000), offers a remarkable theory of the Earth that bears some similarity to Buridan's, though the work was essentially a notebook that set forth Leonardo's ideas about moving water, hydraulics, erosion, springs, and fountains as well as some astronomical topics. The text has been discussed by various authors, and there is a lucid analysis by Gould (1999). Because the manuscript was written in "code," it had no (known) direct influence on subsequent thinking, but it may give an indication of sixteenth-century ideas.

Like others before him, Leonardo supposed that the Earth contained water in great internal chambers. (This idea meshed with the biblical notion of "waters of the deep.") Like Buridan, Leonardo accepted the idea of a geocentric cosmos, with the Earth having more surface water on one side than the other. Additionally, he believed there were scientifically or philosophically significant analogies between "Man," the Earth, and the cosmos as a whole. So, for example, there were supposedly fluids, moving in channels, in human bodies and the "body" of the Earth: both had "veins." And somehow these fluids circulated, though

Figure 2.1: Leonardo's "theory of the Earth" from his "Codex Leicester" (Leonardo da Vinci 2000, 31). © CORBIS/Australian Picture Library. Photographer Seth Joel.

Leonardo did not think of the circulation of the blood as Harvey subsequently conceived it. Thus, Leonardo had the problem of finding a terrestrial analogue to movement of the blood.

He well knew that marine shells were found in strata above sea level. He rejected the "biblical" explanation of their emplacement by the Noachian Flood, pointing to obvious objections, such as that they were not found scattered on the surface, and mollusks moved too slowly to have "walked" to the high ground during the brief period of the Flood. So he needed an explanation for the apparent elevation of the land. His idea was that solid parts of the Earth's interior might sometimes collapse toward the center. A sketch depicting such a collapse is shown in Figure 2.1 (bottom sketch). Such a collapse would shift the Earth's center of gravity. But balance could be restored by the land shown at the top of the sketch *bulging upward.* Thus, fossils—entombed by sedimentation—could be found in elevated strata. The time scale for such processes was unspecified, though I presume that it would have been much less than that required by Buridan's theory. But the theory had the "balance" idea, in common with Buridan.

Leonardo's theory was seemingly not perfectly cyclic, for the Earth would never, by the mechanism proposed, return to some previous configuration exactly. But fragments of the Earth's stratified interior would, one may suppose, collapse to the center from time to time, and the consequent bulging and concomitant deposition of sediment in the oceans would keep the whole in a continually changing but overall steady state, with the center of gravity maintained at the center of the celestial sphere. The internal collapses would be felt on the surface as earthquakes. The upper sketches of Figure 2.1 show how the Earth might appear from the outside ("space") at various times. The elevated land appears in the bottom figure.

Elsewhere in the "Codex," Leonardo revealed his interest in water currents, the erosion of riverbanks, and transport and deposition of sediment. He also entertained the idea that springs might form near the top of hills or mountains by circulation of water within the body of the Earth in a process analogous to those occurring in alchemists' reflux condensers.

As said, Leonardo's ideas were unavailable to the public for many years, and it is unlikely that they had any direct influence on subsequent theorists. Nevertheless, as will be discussed in Chapter 4, ideas were developed in the seventeenth century that showed interesting and perhaps significant parallels to those of Buridan and Leonardo, albeit stripped of Aristotelian notions. To the work of seventeenth-century theorists we now turn.

3

STRAIGHTFORWARD THINKING ABOUT THE EARTH

In Chapter 1, we noted Jaki's distinction between cyclic theories of the ancient world, such as were discussed in Chapter 2, and the linear view of time and history that characterized the Judeo-Christian tradition, and he expressed his sympathies with the latter. With the medieval reconciliation of Christianity and Aristotelian doctrines, the cosmos began to "contract," so to speak. It became a smaller, snugger, world, with the Earth at the center of a sphere of fixed stars (which look closer to us in the clear skies of unpolluted areas of the globe today than they do through the murk of urban smogs). The Moon, Sun, and planets supposedly performed their circuits along their appointed circular trajectories on their celestial spheres, giving resultant motions that were complex to the observer but could be "analyzed" by astronomers' models (all fitted out with a plethora of ad hoc hypotheses). Beyond the stars lay the domain of the various denominations of angels, and above them, and supervising the whole, was the Creator, God, reigning in His heaven and (surprisingly?) having a human face in most medieval and Renaissance representations. The Earth was also thought to be modest in size. In Dante's *Inferno*, the poet supposedly walked through it during an Easter weekend!

The time scale also contracted in the Judeo-Christian imagination. If one took seriously the biblical story of the origin and history of the Earth as described in the Old Testament, then it might be that the Earth and its inhabitants were created in six days, there was only a modest number of generations between Adam and Eve and the birth of Christ, and it was obvious that there were not many centuries between that date and the present. Attempts to calculate the age of the Earth on the basis of biblical history went back to the early Christian era, notably in the work of Theophilus of Antioch (Syria) (ca. 115–ca. 181 A.D.), who in his work *Ad Autolycum* (*To Autolycus*, a polemic intended to refute the views of heathens), based on the "begat" method, affirmed that the Earth was created about 5529 B.C.

(Haber 1959, 17). Such a time scale was refined as years went by, and much scholarly effort went into the task.

The favored date for the Earth's creation that eventually emerged from this work was 4004 B.C., as determined by Archbishop James Ussher (1581–1656) (Ussher 1658), but essentially his result simply depended on the belief that Adam was bound for 4,000 years until redeemed by the coming of Christ. The expected complete "life" of the Earth was 6,000 years, so two-thirds of world history had already elapsed by the birth of Christ! (The extra four years were a "correction" for a sixth-century error in the number of years of the Roman emperor Augustus's reign [Fuller 2005].) Ussher also sought to give dates to the main events described in the Old Testament, and from 1679 his dates for biblical events were given in the margins of copies of the Authorized Version of the King James Bible, thus printing themselves on the public imagination. The strict adherence to the biblical time scale was most prevalent in Protestant countries, where the authority of the Bible, as the "word of God," was thought to outweigh the authority of the Church—a man-made institution.

It was within this context that geological science first began to emerge in the seventeenth century, along with other sciences such as mechanics and anatomy, with advances in scientific instruments such as clocks, microscopes, and telescopes. Rather than adopting a distinction between faith and reason, there was a determined effort to make the new science logically, philosophically, empirically, and theoretically compatible with the biblical accounts of history. Faith and reason were to cohere. So we have what came to be known as "physico-theology," in which religion/theology and the new science mutually supported one another. This worked well for astronomy, even if it entailed the abandonment of the small geocentric Aristotelian cosmos.

Thus, the spatial world picture was radically restructured in the seventeenth century, and the Newtonian synthesis that emerged seemed to offer a grander structure than even the Greeks had envisaged. Not only was it enormously expanded in size, but Newton gave the solar system a mathematically exact description with a predictive capacity that was hard to resist. Indeed, God could be worshipped as the designer of the whole intricate, mechanically functioning cosmos. Aristotle's worldview was discarded, but it had never been essential to Christianity: it was something welded onto it during the Middle Ages.

It was altogether different for time. Christianity appeared to require a short time scale, in accord with scriptural chronology. The history of the world was thought to be known both in the past (from the Old Testament) and, forsooth, into the future (by courtesy of the Book of Revelation). This "knowledge" had to be incorporated in thoughts about the Earth for a successful physico-theology.

Thus, when we reach the seventeenth century, we find theories of the Earth that were specifically intended to provide accounts that accorded with biblical history, which had hardly been the case previously. The most notable example was provided by the English theologian Thomas Burnet (1635–1715). But, before discussing him, we must look at two other figures who had substantial influence on thoughts about the Earth: the French philosopher René Descartes

(1596–1650) and the Danish anatomist, naturalist/geologist, and theologian Nicolaus Steno (1638–1686). In so doing, it will be convenient to distinguish "directionalist" ideas about time and history from the "cyclic" theories that we have mostly considered thus far.

Descartes was a philosopher and mathematician and not much of an experimentalist or observer, though he did some empirical work in optics and anatomy. The ontology of the Aristotelian Middle Ages was that there was a fundamental "substance" or substrate for all things, known by the Greek word *hyle* (Thales' water played an analogous role), on which different "forms" could be impressed (like clay being given various shapes by a potter). The four "elements" had particular "qualities," and all were supposedly inter-convertible. But exactly how this process occurred was unclear. Moreover, on Aristotle's view, there was no empty space in the physical world—no vacuum. There was no such thing as "action at a distance": everything was, in a sense, joined to or interacting with everything else.

There had, however, been an alternative picture, proposed by the Greek philosophers Democritus (ca. 460–437 B.C.) and Leucippus (ca. 430 B.C.): atomism. They supposed that the physical world was made up of indivisible particles (atoms), moving in empty space, and that the different things in the world were formed by different coalitions of atoms, interacting with one another, somewhat in the manner of billiard balls, except that the atoms were too small to be visible and were of various shapes and sizes, too.

In the seventeenth century, this atomistic or corpuscular picture came to be seen as more intelligible than the Aristotelian notions of *hyle,* "forms" and "qualities." Changes from one substance into another were easier to under-stand if seen through the eyes of the atomic theory rather than the vague "forms" and "qualities." Descartes's supposedly "clear and distinct" matter theory was, however, a hybrid of atomism and Aristotelian matter theory. He thought that matter and space *were one and the same.* Space (like numbers) was conceptually and physically unlimited, so the Cartesian material cosmos was infinite. It was supposedly given an initial motion by divine action, but the overall motion could not be linear, for if it were, that would entail leav-ing "empty" space "behind"—which was impossible according to Descartes's suppositions. So he postulated a universe of corpuscles moving in "vortices." And through time, the matter/space would become fragmented into small cor-puscles by the mutual interactions of the constituent parts. Some of the resul-tant particles ("primary") would supposedly be minute spheres, and Descartes thought that these would constitute light (assumed to be a material substance). Smaller "rubbings" ("secondary") would fill the gaps between the spheres, as "heat" or "fire," so that there were no intervening empty spaces. But some rub-bings would coalesce to form variously shaped larger corpuscles ("tertiary"), which would constitute solid earthy substances and air as well.

Continuing his "just-so story," Descartes imagined that there were (for our solar system) initially 14 vortices of different sizes—that which was to become the Sun being the largest. The smaller vortices might have formed "crusts" of tertiary

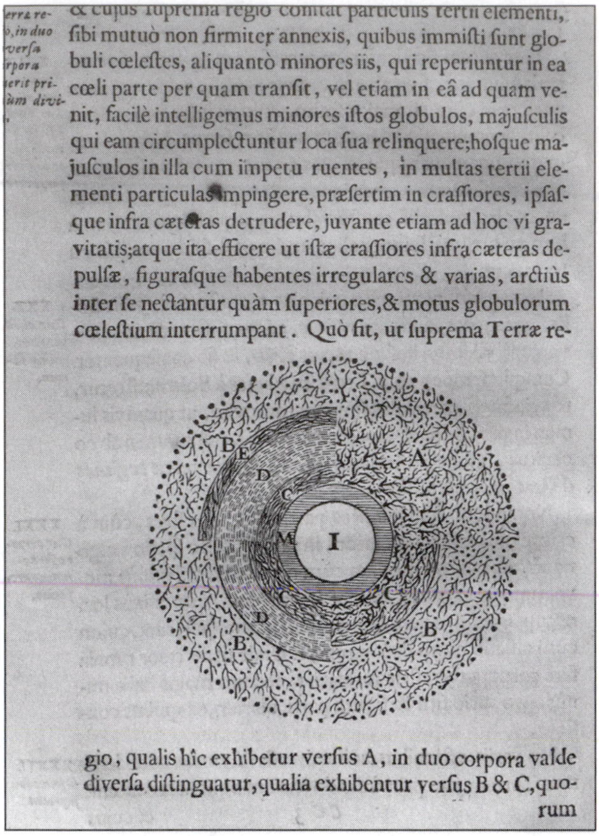

Figure within the figure contains Latin text:

& cujus fuprema regio conftat particulis tertii elementi, fibi mutuò non firmiter annexis, quibus immifti funt globuli cœleftes, aliquantò minores iis, qui reperiuntur in ea cœli parte per quam tranfit, vel etiam in eâ ad quam venit, facilè intelligemus minores iftos globulos, majufculis qui eam circumplectuntur loca fua relinquere;hofque majufculos in illa cum impetu ruentes, in multas tertii elementi particulas impingere, præfertim in craffiores, ipfafque infra cæteras detrudere, juvante etiam ad hoc vi gravitatis;atque ita efficere ut iftæ craffiores infra cæteras depulfæ, figurafque habentes irregulares & varias, arctiùs inter fe nectantùr quàm fuperiores,& motus globulorum cœleftium interrumpant . Quò fit, ut fuprema Terræ re-

gio, qualis hîc exhibetur verfus A, in duo corpora valde diverfa diftinguatur,qualia exhibentur verfus B & C,quorum

Figure 3.1a: Four stages in the Earth's development (Descartes 1644, 206). A and B = atmosphere; C = layer of irregularly branched corpuscles; D = liquid layer (later forming the oceans); E = layer of solid material (the Earth's crust); M = compact layer formed by coalescence of sunspot material; I = region of fiery particles at the Earth's interior. The four quadrants of the diagram represent four stages of the Earth's hypothetical evolution. Image copyright History of Science Collections, University of Oklahoma Libraries.

corpuscles around their fiery interiors and then became caught up in the larger vortices, thus forming satellites orbiting some larger body. Two stages of this process would yield a system of moons, planets, and the Sun (which thus far had formed only incipient crustal material: sunspots). Full details, as envisaged by Descartes for the Earth's formation, need not detain us here, but separate stages are shown in his diagrammatic representation (Figures 3.1a and 3.1b).

This scheme was evidently intended to explain the various astronomical and terrestrial phenomena with which Descartes was familiar. He *claimed* that all his reasoning was (to him) "clear and distinct," and it must be correct, for it yielded a model of the Earth that seemed to accord with the known distribution of land and sea, mountains, and so on (Other matters, such as springs and even the Earth's magnetic field, were also supposedly accounted for.) Whatever the merits or otherwise of the Cartesian model and his logically defective reasoning, it had the advantage, as compared with Aristotle's picture of the cosmos, of conceptual intelligibility as far as the mechanical interactions of the corpuscles were concerned. So the Cartesian "mechanical philosophy" was embraced with no little enthusiasm for more than 50 years, though more in its mechanical, optical, and astronomical aspects than because of its virtues as a theory of the Earth. It will be seen that it had the biblical (or otherwise ancient) idea that there was a large body of water within the Earth. In addition, it was essentially a one-way developmental process, with the Earth perhaps being eventually "captured" by the Sun and becoming part of it. Descartes said nothing about the time scale for the whole process. The vortices notwithstanding, his theory was not cyclic from a geological perspective.

The next person for our consideration, the Danish anatomist, natural philosopher, and theologian Nicolaus Steno (Niels Stensen) (1636–1686), is sometimes regarded as the founder of geology, though James Hutton (Chapter 5) has also been put forward as occupant of this position. Steno's early work was in anatomy, where he made several fundamental discoveries, some of which contradicted Descartes's ideas about the structure of the brain. He traveled extensively and in 1667 became attached to the Medici court in Florence, working with the famous *Accademia del Cimento* (Academy of Experiments) and developing knowledge of the geology of Tuscany. One of his first undertakings in Florence was to dissect the head of a large shark, and this led him to recognize that certain objects found within rocks, popularly called "tongue stones," were in fact sharks' teeth. This led him to ask how they could have gotten into the ground and to the more general question of how a body might come to be contained within another one. On this question, he wrote a small book (intended as an introduction to a larger treatise but not printed, and the manuscript is lost) titled *De solido intra solidum naturaliter contento dissertatione Prodromus* (1669), or *Introduction to a Dissertation on a Solid Naturally Contained within a Solid* (see Steno

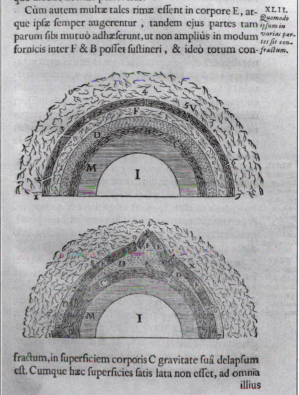

Figure 3.1b: Further development of the Earth (Descartes 1644, 215). This shows collapse structures, with portions of the outer crust falling into the waters beneath and consequent formation of mountains. Descartes supposed that internal cavities had formed by the escape through pores of smaller corpuscles from a lower shell to the Earth's outer surface during hot weather and their failure to return during cold weather. Image copyright History of Science Collections, University of Oklahoma Libraries.

1968). This discussed some general principles that may seem obvious to us but that had never been set down formally so as to facilitate rigorous scientific reasoning about the Earth.

First, Steno asserted that the shells found in strata were indeed the remains of former living organisms. If a thing looked like a shell, it was a shell! Second, the Earth *had a history*, which could be inferred by looking at its present form and structure (Descartes would have agreed) and assuming that the present laws of nature also operated in the past. Moreover, if one body is enclosed in another, the one that hardened first impressed its surface features onto the other one. In addition, if one entity (such as a quartz vein) cuts through another body, then the latter is the older. And, importantly, when sediments are deposited, the lower layers were laid down first or were older. This is the "principle of superposition"—a fundamental assumption of stratigraphy. (Obviously, in a wall, the bottom layer of bricks is laid first and the top one last.) Steno appreciated that strata were laid down horizontally in the first instance but might now sometimes be found inclined to the horizontal. He inferred that departures from the horizontal must have occurred after the deposition, but (perhaps following Descartes?) he was more inclined to entertain collapses of strata rather than elevations due to forces acting from below (or laterally, as we now also think important).

The area where Steno worked—particularly the Arno Valley in Tuscany—seemed to have suffered collapses (at Volterra, for example, large blocks of soft sediments have fallen from the cliffs at the sides of the river toward the riverbed). Steno's reconstruction of the sequence of events seemed to suggested two epochs of sedimentation, two of *under*cutting of strata, and two of collapse. The "history" of the area was depicted diagrammatically in six schematic sections (see Figure 3.2)—the first geological sections ever published. (They have to be "read backward" from 25 to 20.) The first sedimentation (25) could have been associated with a flood at the time of the Earth's creation, and the second one (22) might have been associated with the Noachian flood. Steno's time scale was brief, being of "biblical" dimensions. He knew of the Etruscan burial urns that have been found in the blocks of sediment that have fallen out of the cliffs at Volterra, suggesting a short time scale for the sequence of events. Thus, Steno's geology could accord with the Bible. We might call it a sample of physico-theology, though Steno did not present it in such terms.

Steno's six "cartoons" supposedly showed the history of Tuscany from earliest times to the present. One might therefore think of it as a linear history. But it is better construed as one involving two cycles: similar but also different. As such, there was a prefiguration of an idea that came to the fore in the eighteenth and early nineteenth centuries: a "binary" history, with antediluvian and postdiluvian elements. The two depositions of sedimentary strata (25 and 22) could be ascribed

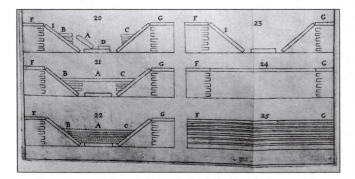

Figure 3.2: Steno's scheme for the geological history of Tuscany, based on the valley of the Arno River and inferences from collapsed strata as there observed (1669, unnumbered page, preceding p. 1). Image copyright History of Science Collections, University of Oklahoma Libraries.

to the events that occurred at the Earth's first creation and then at the time of Noah's Flood. Steno thought that the water for the latter had come from within the Earth, if and when obstructions to some "exits" had been cleared by the expansive action of internally generated hot vapors. Subterranean cavities were also supposedly enlarged or created when the waters retreated whence they came, and collapses subsequently occurred. On this view, the processes were not miraculous but accorded with the operation of the laws of nature. It is hard to admit, however, that the strata in the Arno valley below Volterra look much like 20.

The most remarkable (or egregious?) seventeenth-century effort to propound a theory of the Earth bound to biblical history, as construed by Protestant theologians, was produced by the previously mentioned Reverend Thomas Burnet, Master of Charterhouse and subsequently chaplain to William III, in *Telluris theoria sacra* (1681), or *[Sacred] Theory of the Earth* (Burnet 1684), a work analyzed in detail in Gould (1987). Burnet's work lay outside the bounds of what we would recognize as science but has been seen, nevertheless, as the most popular seventeenth-century book on "geology." It offered an account of Earth history from its divinely created beginning, a transformation from an initial chaos to a smooth paradisiacal planet, fit for the convenient habitation of humans. There followed the unique geological event of the Noachian Flood, which produced a mountainous wreck of a world—all occasioned by mankind's sin. A great conflagration followed at the Day of Judgment, from the dust of which settled a second beauteous world, conducive to the well-being

of the just and godly. This would be reigned over by Christ at his "Second Coming," which would last for a thousand years. The end of the world would come by its transformation to a star, at which point, I suppose, all nonsinners were to go to heaven, the sinners having previously been dispatched by either the Deluge or the great Conflagration. The theory was depicted in the frontispiece to the English edition of Burnet's strange book (Figure 3.3).

It is not hard to see some of the elements that went to make up Burnet's story. The Bible obviously provided the main framework, and one can also see Stoic elements, such as the idea of the Earth undergoing dissolution by water and subsequent destruction by fire. But Burnet also tried to link his story to that of "modern" science, notably that of Descartes and Steno. Sections in the *Theory of Earth* offered by Burnet are reminiscent of Descartes. Compare, for example, Figures 3.4 and 3.1.

The flood itself was supposedly caused by the collapse of the outer

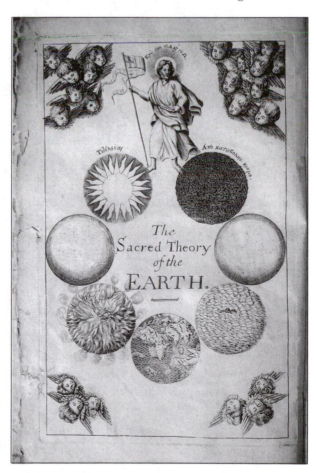

Figure 3.3: Frontispiece to Burnet's *Theory of the Earth* (1684). Image copyright History of Science Collections, University of Oklahoma Libraries.

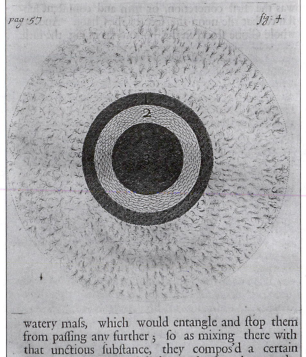

thick, grofs, and dark ; there being an abundance of little Terreftrial particles fwimming in it ftill, after the groffeft were funk down ; which by their heavinefs and lumpifh figure, made their way more eafily and fpeedily. The leffer and lighter which remain'd would fink too, but more flowly, and in a longer time, fo as in their defcent they would meet with that oily liquor upon the face of the Deep, or upon the

pag 57   *fig. 4.*

watery mafs, which would entangle and ftop them from paffing any further ; fo as mixing there with that unctious fubftance, they compos'd a certain

Figure 3.4: Figure from Burnet's *Theory of the Earth* (Burnet 1684, 57). Outer "feathery" cover = air; 1 = Earth's outer crust; 2 = waters of the "Abysse"; 3 = Earth's solid interior. Image copyright History of Science Collections, University of Oklahoma Libraries.

crust into the underlying waters of "the deep," accompanied by the formation of mountains (Figure 3.5). Descartes's influence is evident. Burnet referred to the area of water between *A* and *B* as "the main Chanel [*sic*] of the great Ocean, lying betwixt two Continents."

From one perspective, this was a linear cosmic history: from A (*alpha*) to Ω (*omega*), as represented in Burnet's frontispiece. There was certainly no endless treadmill of cosmic cycles or Stoic increases and decreases of cosmic "tension." Yet Burnet's "just-so story" was a cyclic history—with but *one* cycle. It was a kind of "Ilkley Moor" history! Gould (1987, 41) has generously described it as "the finest expression ever published of the tension between time's arrow and time's cycle." There was indeed a directionalist "arrow." Burnet wrote that he was conducting the history of the Earth "for the space of Seven Thousand Years, through various changes from a *dark chaos to a bright star*," and he likened his story to a stage play. But he also presented his narrative as taking place within the "great circle of time," writing, "When the *Great Year* comes about, with a new order of all things, in the Heavens and on the Earth; and a new dress of nature throughout all her regions . . . [t]his gives a new life to the Creation, and shows the greatness of its Author." I find this inconsistent, for Burnet surely knew that the Great Year, as determined by astronomical observations, was much longer than 7,000 years. Be that as it may (and perhaps the term "Great Year" was not intended to refer to the "Great Year" of astronomers), while he made constant references to biblical texts in his *Sacred Theory*, Burnet believed that he was showing them to be consonant with the best scientific knowledge. However, Cartesian knowledge (of 1644) was becoming obsolete. And publishing his text in 1681, Burnet came just ahead of Newton's *Principia* (1687), which would presumably have made him want to tell a different "story."

Let us return to Steno. His later life was tragic. He became a Catholic and was appointed to a ministry in Protestant northern Germany, where he gave up science and adopted an ascetic "lifestyle," dying in self-imposed poverty. While in Germany, though, he met another philosopher who developed a theory of the Earth, the redoubtable Gottfried Leibniz (1646–1716), mathematician, diplomat, librarian, and court historian, then working for the Duke of Hanover, seeking to write a history of his claimed lands to promote the territorial claims of his House of Brunswick. Both Steno and Leibniz were in Hanover in 1677–1679 and discussed matters of common interest, including geological topics. Leibniz is known to have admired Steno's ideas and to have searched for his lost "geological" manuscript after his death—regrettably without success. It was Steno who prompted Leibniz to produce his own theory of the Earth, his *Protogaea*, or "First Earth." Like Burnet's book, this offered a directionalist Earth history. It was first published as a short note in 1693 (Oldroyd and Howes 1978). Subsequently, it was expanded into a book, published posthumously (Leibniz 1749). A modern edition is also available (Leibniz 1993).

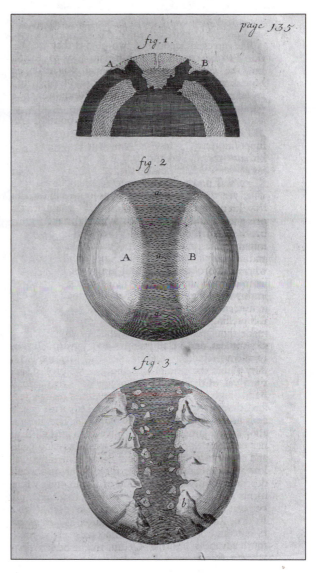

Figure 3.5: Formation of oceans and mountains due to collapse structures (Burnet 1697, 227). Image copyright History of Science Collections, University of Oklahoma Libraries.

Leibniz's theory of the Earth (1693) was somewhat Cartesian, without, however, invoking vortices. He imagined the separation of the heavenly bodies and the Earth from a primeval chaos. They would be luminous or otherwise, according to whether they were or were not covered by an opaque crust. This fire-formed crust would be glassy in character and was subsequently fragmented into sands by the action of condensed water, with the seas becoming salty by materials leached from the solid materials. The oceans formerly covered the high mountains but in time withdrew into subterranean cavities; there might also be crustal collapses into these cavities. It was a "one-way" geohistory.

Such ideas were developed in the full *Protogaea*. Leibniz concurred with Steno that shells embedded in strata were former living organisms, deposited therein along with sediment, and not mere "sports of nature," or creatures that had somehow grown in situ. He also liked Steno's principled way of reasoning about terrestrial phenomena. Leibniz noted the evidence for the sea having covered the area of his employer's dukedom, for there having been volcanoes near the confluence of the Moselle and the Rhine, and for fish formerly having lived near Eisleben. He referred to the Genesis account of the Earth's origin, but it was not regarded as a central issue. He wanted, rather, to do for northern Germany what Steno had done for Tuscany, proposing that there had been *successive* inundations, though these eventually led to a stabilized condition (Leibniz 1993, 21, 23). Leibniz contemplated but rejected the idea (from Buridan?) that geological changes (inundations and so on) might be caused by shifts of the Earth's center of gravity (Leibniz 1993, 29), for—perhaps considering the Americas—it was, by Leibniz's time, less plausible to suppose that there were distinct ocean and land "sides" to the Earth, as supposed by Buridan.

Leibniz's theory was more "arrow-like" than cyclic (which shows that not all geotheories need be cyclic). His work is interesting to us in that it provides a good example of someone seeking to invoke both fire and water as geological agents, with water rounding pebbles, eroding valleys, and depositing fossiliferous sediments. His ideas on subterranean winds were archaic, but the notion of the Earth having a central reservoir of heat remains important to the present.

The last part of *Protogaea* gave information from the sinking of wells at Modena and Rosdorf (near Göttingen), where the strata seemed to suggest "successive inundations" (Leibniz 1993, 155). Additionally, a profile of strata observed in a well near Amsterdam was tabulated (Leibniz 1993, 197) based, according to Yamada (2003: 94), on information found in Varenius's *Geographia generalis* of 1650.

Garden soil	7 feet
Peat	9
Clay	9
Sand	8
Earth	4
Clay	10
Earth	4
Sand, on which are built the foundation piles of the houses of Amsterdam	10
Clay	2
White sand	4
Dry earth	5
Disordered materials	1
Sand	14
Sandy clay	3

Sands mixed with clay	5
Sands containing marine shells	4
Clay	200
Coarse sand	31

Such data would have done credit to a nineteenth-century stratigrapher, measuring up a profile. It is relevant here in that it provides evidence for knowledge of the alternations of strata, which Leibniz thought significant, even if they did not show clear manifestations of cyclic deposition. Sometimes the world almost spoke in terms of recurrence—even to the "directionalist" Leibniz.

4

ROBERT HOOKE AND THE FIRST TESTABLE-IN-PRINCIPLE THEORY OF GEOLOGICAL CYCLES

We now retrace our steps to examine the work of the English polymath Robert Hooke (1635–1702). I deal with him after Burnet and Leibniz, for his ideas about the Earth were in advance of either of them, though earlier, and were intended to have *experimental* support as well as accounting for field observations. Hooke used the more advanced scientific information of the later seventeenth century, with its improved understanding of mechanics, and astronomical instruments that magnified the appearances of objects in the sky. However, in his later work he began to use unreliable sources like Ovid to support his geological hypotheses.

Hooke came from the Isle of Wight, off the southern coast of England, where strata are well exposed, especially chalk. Gigantic fossil ammonites could be found at Portland, to the west. Hooke was an inventor (especially of scientific instruments), astronomer, physicist, microscopist, musician, philosopher of science, mathematician, surveyor, architect, and much besides. He was well read, especially in the classics, as his library sale catalog reveals (Smith 1703). Hooke was not well off initially and began his career as assistant to the Honourable Robert Boyle (of Boyle's law fame) at Oxford. Not long after, Hooke became curator of experiments to the fledgling Royal Society, where he was expected to produce experiments at its meetings for the edification of the members, and soon he was the Society's secretary. It was largely due to Hooke that the Society survived. He eventually became quite rich as a result of his work on resurveying London after the Great Fire (1666) and his designs for buildings thereafter. He was jealous of his reputation and became involved in various disputes for which he acquired a reputation of being cantankerous, but recent researches have questioned this characterization. His health was never good and deteriorated over the years, so that he was probably much in pain and hence irascible. He became geometry professor at Gresham College, a body established in London by an Elizabethan merchant, Thomas Gresham, where lectures were presented in various fields.

Hooke had an abiding interest in geological matters and gave lectures on his observations and theories through much of his career (Rappaport 1986). They were published only posthumously (Waller 1705). The first series was delivered between June 1667 and September 1668. The second series, presented in 1686 and 1687, contained materials on philosophy of science as well as a novel "theory of the Earth." They provoked opposition, being incompatible with the biblical time scale (Oldroyd 1989; Turner 1974). Hooke sought to refute the objections in a series of lectures from late 1687 by offering geological interpretations of events described by such ancient authors as Plato and Ovid (Birkett and Oldroyd 1991). Thereafter, Hooke continued to bring geological matters to the Royal Society's attention, chiefly adducing mythopoeic evidence.

Hooke's lectures were prolix but contained matters of great significance in the history of geoscience. The first theme was the occurrence of fossils in strata and arguments that they were the petrified remains of former organisms rather than "sports of nature" or the product of some mysterious "Plastick or Vegetative Faculty working in Stones." As Ellen Drake (1996) has emphasized, much of Hooke's discussion was based on his observations in the Isle of Wight, where he had found, among other things, evidences of erosion and the washing out of fossils from its chalk cliffs. The published version of the "Discourse" was accompanied by beautiful drawings of fossils, especially ammonites ("snake stones"). Hooke was a notable observer and draftsman.

Hooke's evidence suggested, then, that there had been interchanges of the relative levels of land and sea—and not just due to Noah's Flood. He thought that the changes might be associated with earthquakes (which he later linked with volcanoes). Moreover, it was "not unlikely also but that there . . . [might] be divers [i.e., several] new kinds [of organisms] now, which have not been from the beginning" (Hooke 1705, 291), arising from environmental changes (Hooke 1705, 327). Soundings suggested that there were inequalities on the bottom of the sea, analogous to plains and mountains on land. Hooke also adumbrated the idea of cycles of erosion, deposition, and consolidation of earth, just as there are astronomical cycles, or life (and death) cycles (Hooke 1705, 312–13). At this stage, he offered no explanation of elevation other than it could sometimes be caused by earthquakes. Yet he went on to suggest that volcanoes emitting material at one place might lead to subsidences elsewhere, thus altering the Earth's form (Hooke 1705, 321) and presumably its center of gravity. In addition, a great earthquake might alter the position of the Earth's center of gravity or its poles of rotation, and the Earth's rate of rotation might have decreased over time because of the effect of "the fluid Medium in which it moves" (Hooke 1705, 322). The force of gravitational attraction itself might change. Hooke further speculated that the Earth might be aging, despite cycles of renewal (see the discussion later in this chapter), as its presumed internal heat became exhausted. In considering global changes of latitudes, meridians, pole points, or gravitational attraction, he was thinking of analogies with known changes in the Earth's magnetism (magnetic field as we would say) or the fact that if a fragment is broken off a lodestone (a magnetic oxide of iron) and placed on it elsewhere, this could cause a change in the position of its magnetic poles. I agree with Hooke that changes of the distribution of the Earth's mass (i.e., changes in the position of its center of gravity), whether caused by erosion and deposition, volcanic activity, structural collapses, or whatever, could (in principle) alter the alignment of the polar axis with respect to the Earth's surface.

Hooke's lectures of 1686–1687 started with a discussion of philosophy of science, which need not be detailed here beyond saying that his recommended procedure in science was as follows: make observations of phenomena that are puzzling and require explanation, propound explanatory hypotheses, deduce possible testable consequences of each hypothesis, carry out tests to try to determine which is the most satisfactory hypothesis in the light of the results, and use the information thus generated as the basis for further hypothesizing and testing. This was the so-called hypothetico-deductive procedure in science. It is remarkable that it first found such a clear exposition and exemplification in geological work.

Hooke then went over his old arguments about fossils, erosion, and deposition and changing levels of land and sea, such as he had given in his lectures some 20 years previously. He also reaffirmed the hypothesis of pole wandering (cf. Buridan/Oresme) but in a new way and with new implications:

I suppose . . . that the *Axis* of the *Diurnal Rotation* of the Earth ha[th] . . . had a progressive motion, and hath, in process of time, been chang'd in position within the Body of the Earth, and consequently that the Poler [*sic*] Points upon the Surface of the Earth, have alter'd their Situation; so that the present Polar Points have formerly been [are now?] distant from those Poles that were then; and consequently that those former Polar Points are now remov'd to a certain distance from the present, and move in Circles about the present. (Hooke 1705, 346)

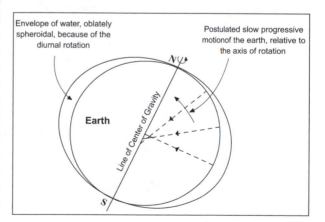

Envelope of water, oblately spheroidal, because of the diurnal rotation

Postulated slow progressive motion of the earth, relative to the axis of rotation

Earth

Line of Center of Gravity

Further, Hooke correctly supposed that the Earth was an oblate spheroid (flattened at the poles), in

Figure 4.1: The Earth, its envelope of water, and the initial directions of motion of freely falling objects at different latitudes, based on Hooke's account in his *Posthumous Works*. Reproduced from *British Journal for the History of Science* (1972, 126) by permission of Cambridge University Press. Redrawn by Ricochet Productions.

contrast with the Cartesians, who held that it was a prolate spheroid. Hooke correctly stated that the envelope of the Earth's waters was more spheroidal than the Earth itself. He also assumed that the force of gravitational attraction was everywhere directed at right angles to the Earth's surface. So, given that it was not perfectly spherical, it would follow that gravitational attraction *at the surface* would be directed not to the Earth's geometrical center but to some point on a line joining the two poles. As Hooke put it, the Earth's center of gravity could be "drawn out into a line into the *Axis* thereof." His suggestion is represented in Figure 4.1.

The Earth itself had, of course, many inequalities, with high ground and low. So land that is underwater at the equator might become exposed to the atmosphere if, by pole wandering, it moved toward a region of higher latitude, and it would then be weathered and eroded. Conversely, land would become inundated and covered by sediments as it approached the equatorial regions. Hence,

> [T]he Earth itself doth, as it were, wash and smooth its own Face, and by degrees to remove all the Warts, Furrows, Wrinckles and Holes of her Skin, which Age and Distempers [diseases or bodily disorders] have produced. (Hooke 1705, 348)

Thus, he suggested a genuinely cyclic theory of erosion and deposition. Moreover, as a given locality moved toward or away from a pole, the direction of the gravitational attraction at a point on the surface would change. Additionally, the rapid motion of points on the equator or other latitudes, as compared with a pole, would produce different overall forces. (Hooke mentioned that one had to shorten a pendulum slightly in the tropics to make it keep "due Time.") So the slow changes of forces associated with the hypothesized pole wandering could cause occasional sudden adjustments to the forces—or earthquakes.

Hooke emphasized that he was simply proposing a *hypothesis* that, if true, would account for some important geological observations. As far as I'm aware, his idea was largely original, though he may have obtained the idea of pole wandering from the Buridan tradition. Hooke was well read, but his library had few books from the Middle Ages, which were largely unavailable in Hooke's day, whereas the classic works of Ovid and so on were available in "modern" editions. It should be noted that Hooke was *not* postulating continental drift, which entails movement of different parts of the Earth's crust relative to one another and to the material of the Earth's interior. Nor was he suggesting changes in the obliquity of the ecliptic (the angle between the Earth's axis of rotation and the plane defined by the Earth's orbit around the Sun).

The theory then needed testing. Hooke never accomplished this successfully, but he did make suggestions as to what might be done using long-term astronomical observations. His argument was that if the positions of the poles and the latitudes of particular places on the Earth's surface are changing, then there should be slow changes in the directions of the meridians (north–south directions) at different places. So *if* the direction of a meridian were determined accurately at some place and *if* the determination were repeated there after some substantial interval of time, *then* a detected shift in the meridional direction would support Hooke's theory. But he had no idea how long he might have to wait to find some change of meridional direction or how accurate his instruments needed to be.

Nevertheless, Hooke proposed two methods for determining a meridional direction accurately. The first entailed the use of a long telescope with the objective lens on the top of a building and the eyepiece near the ground. The pole star should be observed on some date and its most easterly and westerly positions determined during the night. The pole of the heavens would be the position halfway between these two points. Using a plumb line suspended from near the telescope's objective lens when it was directed exactly toward the pole, the meridian direction could then be determined on the ground. If the telescope was then directed horizontally along the line of the meridian, one could draw what was observed in the field of view (some building or whatever). By repetition of the determination some years later, one might ascertain whether there was a change in the field of view resulting from a change in the direction of the meridian.

For his second method, Hooke proposed that a series of concentric circles be inscribed on a piece of glass placed between the eyepiece and the objective. The telescope was then to be directed toward the North Pole, where Hooke had discovered a small constellation, invisible to the naked eye, that he named the English Rose. When the telescope was pointing *exactly* toward the pole of the heavens, the stars of this constellation would describe concentric circles during a night, as judged by the lines on the glass. The line of meridian could then be found by the use of two plumb lines suspended from the pole-pointing telescope to the ground. Again one should come back some years later and repeat the determination. An attempt to make a determination of the meridian by Hooke's telescopic method was made on March 9, 1687 (Birch 1756–1757, 4: 527), but it failed because of bad weather, and the matter was not pursued.

Hooke's "geotheory" was discussed early in 1687 by some academics at Oxford, headed by the mathematician John Wallis. It was thought unacceptable, as it was incompatible with what was supposedly known about the Earth's history (with its short "biblical" time scale) and about topographic changes that had been recorded in history. The criticisms were conveyed to Hooke by the astronomer Edmund Halley, and an acrimonious correspondence between Hooke and Wallis ensued. It was in the course of this exchange that Hooke first proposed his ideas about the determination of meridians (Oldroyd 1989).

Hooke claimed that the Danish mathematician and natural philosopher Erasmus Bartholin had published some evidence for meridional changes in 1673, but a report by the German astronomer Johann Philipp Wurtzelbauer in the Royal Society's *Philosophical Transactions* in November 1687 claimed that there had been no measurable changes in the meridional direction at Nurenburg during the previous 200 years. So Hooke put aside his suggestions, and in his later addresses to the Royal Society, he focused on possible evidences for geological changes that might be inferred from ancient texts. This approach may be called euhemerism—a term named after the Sicilian author Euhemeris (fl. 316 B.C.), who thought that the gods of Greek mythology were *real* people who had been deified or myths derived from events that actually occurred. I like the term "physico-mythology," in which science and myth are used to mutually reinforce one another.

In lectures from 1687 to 1700, Hooke examined accounts from Hanno the Carthaginian (who is believed to have explored the northwestern coast of Africa), Plato, Virgil, Ovid, Pliny, and Herodotus as well as the Bible (Birkett and Oldroyd 1991). We need not pursue Hooke's interpretations here. But as an example, I mention Ovid's description of Perseus's rescue of the maiden Andromeda, chained to a rock and about to be devoured by a sea monster. Perseus (whom Hooke took to be a personification of lightning) saw the maiden's hair waving, fell in love with her, and attacked and killed the monster. Hooke imagined that the event occurred in North Africa, where there was rocky country that was alternately covered by high tides and by sands. He construed Andromeda's name to mean "raised from inundation," and the events referred to elevation of the North African coast. The waving hair might refer to waving reeds. He was drawing a long bow!

So Hooke's late geological thinking was designed to integrate information from ancient sources that might provide a warrant for his earlier cyclic hypotheses. His time scale was evidently contracting. He imagined the Earth to have been formed, at the Creation, of a central solid core (as we might say), a surrounding spherical shell of water (the Waters of the Deep), and another one of solid material (more or less equivalent to our crust), which he called the "firmament," which in turn was overlain by the oceans. (The term "firmament" more commonly refers to the celestial sphere of the sky.) Hooke suggested, then, that, according to *Genesis* (verse 9), the Earth's outer crust became deformed, so that it was no longer spherical; the seas were gathered into basins; and land elevation occurred: "God said: Let the waters under the heaven be gathered together upon one place, and let the dry land appear." In his old age, Hooke had become a scriptural geologist, akin to Burnet! So the brilliant early theorist became a tedious physico-theologian/physico-mythologist.

Perhaps for this reason, Hooke's geological work was largely forgotten for nearly a century, his reputation becoming overshadowed by Newton's mathematics, optics, astronomy, and theoretical mechanics, with his commanding position as president of the Royal Society. But as we shall see (Chapter 5), Hooke's geology was not entirely forgotten and was deployed to some extent by James Hutton, who, with his cyclic theorizing, is often represented as the founder of modern geology.

5

THE FOUNDER OF GEOLOGY? JAMES HUTTON, THE GEOSTROPHIC CYCLE, AND A FRENCH "CYCLIST"

We now consider perhaps the best-known "cyclist" in the history of geology, often regarded as the "Founder of Geology" (Dean 1992, 1), and it was his efforts as a cyclist that had much to do with his high standing in geoscience history.

James Hutton (1726–1797) was born in Edinburgh, son of a businessman. Hutton Senior died, however, when James was only three, bequeathing him two farms in Berwickshire, one of the better areas for Scottish agriculture. The young Hutton attended Edinburgh High School and at the tender age of 14 went on to the university to study humanities, which enabled him to attend the lectures of the mathematician Colin Maclaurin, a former protégé of Isaac Newton. Hutton also acquired an interest in chemistry. After graduation, he was apprenticed at a lawyer's office but found the work uncongenial, so he turned to medicine, probably hoping to link the emerging science of chemistry with an income-yielding career.

Although Edinburgh was the leading university for medicine in Britain, after completing the course there Hutton went to Paris in 1747 for further study, where he became fluent in French. He then moved to Leiden, where the celebrated medico Hermann Boerhaave (1668–1738) taught and where Hutton presented a doctoral thesis in 1749. It was concerned with the *circulation* of the blood. It also gave much attention to chemical matters and provided a good conspectus of the anatomy and physiology of the human body, as known in the mid-eighteenth century. The thesis contained the following portentous paragraph (originally in Latin):

Being about to treat of blood and bodily fluids, I shall, as far as possible, ignore solid materials in order to avoid digressions from the main subject-matter, although there exists so close a connection between those related items which mutually give form to and modify one another that it is not easy to determine which of them has prior existence, both advancing in step, one refashioning the other and one modifying the other; and thus they display the glorious cycle of life and a very beautiful instance of a perpetual motion machine—an instance in which matter moves without a material cause, in which it seeks its own special aims on the fertile earth, and in which it reconstitutes its daily diminutions by means of the very cause of its destruction; and before yielding its life-producing movement to the fatal necessity of material machines, it produces new offshoots, which will repeat its role in the microcosmic grove. (Donovan and Prentiss 1980, 30)

Here, Hutton used the word "microcosm" to refer to the human body, which, post-Harvey, was known to employ cyclic processes. As previously mentioned, in medieval and Renaissance thought, the human body ("microcosm") was imagined to have analogies with the Earth or to the whole cosmos ("macrocosm"). So even at the beginning of his career, in his use of old-fashioned diction, Hutton was seemingly prepared to contemplate cyclic processes in or on the Earth, analogous to those that occurred in the body.

Returning to Britain, Hutton was initially in London and then in Edinburgh for two years. He did not, however, take up medicine but went into partnership with a friend, James Davie, with a process for extracting "sal ammoniac" (ammonium chloride) from soot (of which there was plenty in Edinburgh!). Essentially, the sal ammoniac was separated by sublimation. It must have been a filthy process, even by eighteenth-century standards. However, the business prospered and eventually gave the partners a steady income. But Hutton's personal life fell into disarray with the birth of a son, probably out of wedlock, and he left Edinburgh for several years and began a farming career. He retired to his properties in Berwickshire, but before settling down there, he went to East Anglia to study the latest agricultural methods. Later he brought such techniques north to Scotland. During his two years away, Hutton traveled extensively, and, becoming increasingly interested in the Earth, he recognized the endless action of the forces of erosion and saw sedimentary rocks for what they were: consolidated sediments.

Back in Scotland, Hutton had to do some of the farm work himself. He was not (initially) just a gentleman-farmer or absentee landlord, and he experimented in his agricultural practice in various ways. Under Maclaurin, Hutton had become acquainted with "deism," which was in a sense a development of the attempted rationalism of physico-theology, but shorn of reference to the Bible, which was now (again "in a sense") replaced by Newtonian mechanics and astronomy. Deists regarded God as the designer of the cosmos, who had prepared it for the well-being of mankind, furnishing the Earth with the necessities of air, water, and soil. The deists' God was not a personal entity. *It* did not interest itself in our moral triumphs or failures. Jesus was not a divine being. So deism was not congruent with Christianity. For some, however, it was the theology of eighteenth-century Enlightenment and also the thin end of the wedge of atheism. Scientific evidence supposedly demonstrated the existence of God, by courtesy of the "argument from design." For example, the "design" of the parts of animals or plants as functioning wholes was so remarkable that it could not have happened "by chance." There *had* to be a designer, responsible for it all. It appears that Hutton lost his Christian faith at some early stage of his life, but it was replaced by an enthusiastic deism.

Farmer Hutton was struck by the fact that soil was constantly washing into the sea, and, as soil was essential for human well-being, he thought that it had to be replenished, to be consistent with divine design. Indeed, he could see that it *was* being replenished, by weathering, erosion, and transport. Hutton did not believe in a biblical time scale. The Earth was, he thought, for all practical purposes, infinitely old. But were that so, the land would eventually be eroded to a plain, without hills or mountains, and the source of fresh soil would be gone. So the high ground had to be replenished, to allow the continued formation of new soil. How? This was the problem of land elevation, which, as we've seen, had been exercising people's attention for centuries.

In 1764, Hutton journeyed into the Highlands with a gentleman friend, George Clerk-Maxwell, and began the systematic collection of geological information and specimens. He persevered with his farming and eventually got the better of his land and began to make a profit. The sal ammoniac business also prospered, and Hutton began to participate in the activities of the Industrial Revolution, such as the construction of a canal between Glasgow and Edinburgh. About 1767, he rented out his farms and returned to Edinburgh (his old scandal having subsided), with an income to enable him to enjoy the pleasures of intellectual life in the Scottish City of the Enlightenment. He was gregarious and clubbable, though some of his correspondence suggests he was a lewd, even lecherous man, perhaps because of his long bucolic sojourn in Berwickshire.

Among Hutton's new friends were the economist Adam Smith, the chemist Joseph Black (who conducted experiments on questions to do with heat), and the famous engineer, instrument maker, and inventor James Watt. It was Watt's steam engine that led Hutton to think of heat as an agent of geological change and the possibility of the Earth having a central source of heat that might drive the cycle that he envisaged in his theory of the Earth. Perhaps heat could both consolidate sediments and elevate strata? Terrestrial heat could have a function analogous to that of the boiler fire of Watt's engines, which drove their heavy and complicated mechanisms.

In 1774, Hutton and Watt visited a factory in Birmingham, where a Watt engine was in operation. Following this visit, Hutton traveled into Wales and the west of England, collecting geological specimens and information, then back into the Midlands and off to Wales again, as far as Anglesey. After visit further interesting localities in the Midlands, he headed back to Edinburgh, where he was soon diversifying his industrial interests, working on the production of varnishes.

Hutton's geotheory was worked out in general terms by the 1780s and probably earlier according to his biographer John Playfair (1805, 55–56). It was as follows. The Earth, Hutton thought, has a central heat reservoir, the origin or source of which, or its mode of maintenance, was unspecified (but it wasn't due to the combustion of flammable substances). Rocks on land are changed by weathering and erosion to form soils. Sediments are deposited in the seas by rivers, which also carve valleys, and accumulate in layers on the sea floors (cf. Steno). The lower layers are compressed by the sediments deposited on top of them and in time become consolidated, assisted by the Earth's internal heat (the lower sediments being nearer to the internal heat reservoir than those overlying them or materials on land).

Hutton knew that gases become hotter when compressed (as a bicycle pump becomes heated when in use) or cooled when they expand, and possibly knowledge of such phenomena lay at the bottom of his theory. In time, the consolidated materials, under pressure, might become so hot that they melt, expand, and then push upward into the Earth's crust, passing through joints or other lines of weakness. Thus, veins of crystalline rock might be found in the rock, apparently (from their observed geometry and configurations) having moved upward from below rather than downward from above. They could form light-colored granitic or mineral veins. (But Hutton didn't have firsthand knowledge of granitic veins when he initially developed his theory.) Or there could be intruded material that resembled lava from volcanoes. The latter—generally dark, crystalline, masses—could sometimes cut across strata (and are thus younger, according to Steno's principle) to form "dykes" (or walls). Or they might penetrate between the layers of strata, in which case they were called "sills." Hutton knew examples of both kinds of intrusive structure on the hill behind his Edinburgh home.

On a grander scale, Hutton supposed that great masses of molten material (which we call magma) could be intruded into the Earth's crust, heaving it up into dome-like structures. On cooling nearer the surface, the magma crystallized to form subterranean granite masses, which might subsequently be exposed by erosion, and may now be seen in the cores of mountain ranges. Thus, the land would be renewed, and Hutton's Earth ("designed in wisdom") could continue indefinitely as a place suited to human habitation. The whole process was cyclic.

Not only that, the upheaved strata might be planed down by erosion to form new surfaces on which additional sediments might subsequently be deposited horizontally. So on looking at sections of strata today, one might find places where the lower layers are inclined to the horizontal and the overlying ones overlie them horizontally. Such a structure came to be known as an *unconformity*, and Hutton's later discovery of unconformities was considered a triumph for his theory. His theory allowed him to make a successful *prediction*.

The theory was profoundly cyclic, and the geohistory that it envisaged was open ended. Hutton did not assert that the Earth was infinitely old, but, as he put it in a famous sentence, "we find no vestige of a beginning,—no prospect of an end" (Hutton 1788, 304). And the sequences of formation and emplacement of magma, the erosion of land, and the deposition and consolidation of sediments required an enormous age for the Earth.

The cycle thus envisaged has been called the "geostrophic cycle" by Tomkeieff (1948, 1962), who represented it as shown in Figure 5.1, which illustrates the cyclic nature of Huttonian theory and the concept of unconformity.

Hutton's theory was eventually presented in a paper read to the Royal Society of Edinburgh in 1785 (Hutton 1788; abridged in Oldroyd 2000) and in expanded form in his two-volume *Theory of the Earth* (Hutton 1795). A third volume remained unpublished in his lifetime, but the manuscript was found in the nineteenth century and was published by Archibald Geikie (Hutton 1899). This third volume is important, as it described field excursions that Hutton undertook after the initial presentation of his paper in 1785 and that provided empirical support for his ideas.

Hutton's 1785 (1788) paper did not explain the Earth's internal heat. Rather, it sought to use field specimens to substantiate its existence. Many of the materials that bind sediments together, such as calcareous spar, silica, and so on, are not themselves readily water soluble. But heat could penetrate into bodies and, by fusion of their particles, followed by cooling, could cause consolidation. So (he wrongly thought) heat must have been responsible for the penetration of sediments by flint nodules (injected while molten!): they could not, he maintained, have been deposited by water, as they are not water soluble. Likewise, nodules containing crystalline spar, which doesn't extend to their outer surfaces, could not (he thought) have acquired the spar matter by transport of solutions. A sectioned specimen of such a nodule was presented as supposed evidence for the agency of heat in the 1785/1788 paper.

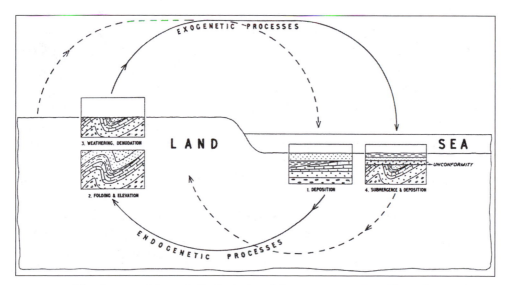

Figure 5.1: The "geostrophic cycle." Reproduced by permission from *Proceedings of The Geologists' Association,* S.I. Tomkeieff, "Unconformity—An Historical Study." 73: 396, fig. 6 © 1962 The Geologists' Association. (The term "geostrophic" is used in oceanography to describe a situation in which a wind or an oceanic current is balanced by an equal and opposite Coriolis force.)

Hutton also acquired a specimen of granite from Portsoy, near Aberdeen, that contained crystals of quartz within feldspar within quartz. It seemed implausible that such an arrangement could have been produced by crystallization from an aqueous solution, as was supposedly the case for granite according to contemporary German theorists, notably Abraham Gottlob Werner (1749–1817) from the Freiberg Mining Academy, whose "straightforward" (directionalist), "aqueous" theory was being taught and well received on the Continent.[1]

As said, Hutton's theory was paradigmatically cyclic. It was almost of the "captain and mate" variety. But there were gaps in the story. As in the case of Harvey, who reckoned that their *had* to be capillaries linking the arteries and the veins, to make his experiments intelligible, and to make possible the movement of blood around the body, so, Hutton argued, there *had* to be heat within the Earth, and there *had* to be some means of elevation, even if he did not know precisely how that process worked. The hot interior was evidenced by volcanoes of course, and mines seemed to have higher temperatures at greater depths—but it was not until the nineteenth century that detailed investigations of mine temperatures confirmed this (e.g., Cordier 1827).

So while of fundamental importance, Hutton's 1785/1788 paper lacked important supporting evidence. In 1785, he hadn't even recorded observations of granite veins penetrating other rocks, and there's no documented reason to believe he had discovered any unconformities. But subsequently, Hutton made fresh expeditions and found the sorts of things he predicted.

In September 1785, Hutton went into the Scottish mountains to hunt for and examine contacts between granite and the surrounding rocks ("country rock") into which it had supposedly been injected while molten. He was accompanied by his friend John Clerk of Eldin, an accomplished artist, who made excellent drawings of what they saw (Craig, McIntyre, and Waterston 1978). I don't know how Hutton knew where to go, though it is likely that he thought he would find the evidence he wanted to the west of the well-known mass of Aberdeen granite. He must, in fact, have received some specific hints, for he and Clerk headed directly for an outcrop in the valley of the River Tilt in central Scotland, which runs northeast from Blair Atholl, seat of the Duke of Atholl. Complicated outcrops of limestones and schists were found in the valley, and not far up the glen they came across a wondrous set of granitic veins, both wide and narrow, that sometimes cut across the country rock while at other spots the veins could be seen anastomosing between or through the laminae of the "country rock." It was recorded by Playfair (1805, 68–69) that Hutton was so excited by what he saw that the local guides imagined that he'd found "nothing less than . . . a vein of silver or gold, [for only] that could [have] call[ed] forth such strong marks of joy and exultation"! On further examination, they were able to trace the granitic veins to the large mass of granite to the north of the glen. Hutton's joy was, of course, due to the fact that he had found what he had *predicted* on the basis of his theory, and the geometry of the veins was compatible only with the granite having worked its way into the country rock from below from subterranean granite magma rather than being precipitated from above, as envisaged by Werner's rival theory (see p. 57).

In 1787, Hutton visited what is now a classic site in the history of geology: the Isle of Arran, west of Glasgow. This has a massive outcrop of granite, with surrounding tilted-up layers of schist and beds of sandstones and other sediments stratigraphically above the schists (but at lower altitude). A block of schist, traversed by granite veins, was brought back to Edinburgh to convince skeptics of the virtues of Hutton's theory. Clerk's son (another John) was with Hutton and produced a wonderful hypothetical cross section of the island, construed in terms of Hutton's theory (Figure 5.2). The section is compatible with modern geological maps of the area, which can be interpreted as representing a domed structure of schists and sedimentary rocks disposed around a central granite core (Figure 5.3).

In 1786, Hutton made trips into southwestern Scotland to look for further granite masses and to examine their contacts with the country rocks. He reported his observations as being most satisfactory for his theory (Hutton 1794b). He argued from the geometrical relationship of the granitic veins that they were emplaced *after* the country rock. So, contrary to Werner's theory, granite was not an "original," "primitive," or "primeval" rock. Hutton acknowledged, however, that some of the granites in northern Scotland had a stratified or foliated appearances. So we needn't be surprised that controversies about the origin of granite persisted until the twentieth century or that the Glen Tilt observations did not provide the last word on the topic. There are granites and granites! (Some may have been metamorphosed so as to yield a foliated structure. Others may be generated by rocks being impregnated with hot mobile fluids and not by the straightforward cooling of large masses of magma.)

Figure 5.2: Cross section of northern part of Arran, drawn by John Clerk Jr., 1787. © Sir Robert Clerk. Reproduced by permission.

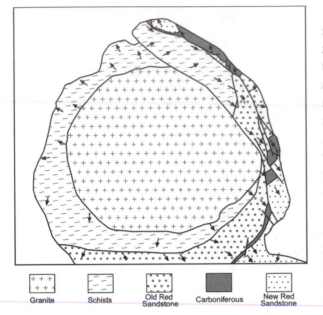

Granite   Schists   Old Red Sandstone   Carboniferous   New Red Sandstone

Figure 5.3: Geological sketch map of northern Arran, showing likely line of section for Clerk's drawing (Figure 5.2). Arrows indicate the dips of layered strata. Reproduced by permission from D. Oldroyd, "History of Geology from 1785 to 1835," in R. Selley et al., eds., *Encyclopedia of Geology* (Oxford: Elsevier, 2005), 3:175. © 2005 Elsevier.

Also at the northern tip of Arran, near Loch Ranza, Hutton found his first evidence of unconformity, with the sedimentary strata (sandstones and limestones) lying over inclined schists (Figure 5.4). Again he had made a successful prediction. A "swarm" of dark basaltic dykes was also observed, cutting through the low-lying beds by Arran's southern shore.

The Loch Ranza exposure was not, however, as convincing as Hutton might have wished (or as Geikie's sketch suggests), for the rocks were obscured by vegetation. But on their way back to Edinburgh, the travelers found an excellent unconformity near Jedburgh in the southern border area of Scotland, where the road ran by the River Jed's banks and a section revealed a fine view of Old Red Sandstone, lying horizontally over upended gray, gritty sandstone (greywacke)—which Hutton called "schistus," though it was not what is today called a schist (a metamorphic rock) (Figure 5.5). Just below the horizontal sandstones, Hutton figured what we call a "basal conglomerate." It contained fragments of the underlying "schistus" embedded in a matrix of the material of the overlying sandstone. This indicated that erosion of the "schistus" occurred before the deposition of the sandstone sediments. It was real progress for Hutton's theory. We also see that veins of igneous rock (not so called) were drawn by Clerk, penetrating into and disturbing the layers of the greywacke. The figuration of the strata in this illustration will attract our later attention (p. 109).

But the most famous examination of an unconformity occurred in June 1788. Hutton's upland farm was located on "schistus," while his lowland farm was on soil derived from Old Red Sandstone. The boundary between these two rock types ran northward to the coast, approximately along the eastern boundary of the upland farm. Hutton must have been well aware of the two rock types, similar to those he had seen at Jedburgh. It seemed a good

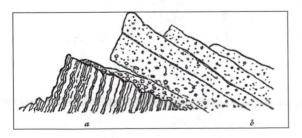

Figure 5.4: Unconformity at Loch Ranza, Arran, as figured by Archibald Geikie (1899).

Figure 5.5: Profile of unconformity observed by Hutton near Jedburgh in 1787. Hutton (1795, vol. 1, pl. III). Image copyright History of Science Collections, University of Oklahoma Libraries.

plan to examine the contact at the coast, where an exposed contact might be found. Accordingly, with two friends (the mathematician, John Playfair, and a local landowner, Sir James Hall), he sailed eastward along the Berwickshire coast, passing the terrain of schistus and reaching that of sandstone. To their great satisfaction, they encountered a superb unconformable contact at a place called "Siccar Point," with Old Red Sandstone overlying the schistus, as at Jedburgh, but exposed in such a way that the three-dimensional structure of the contact was evident (Figure 5.6).

The Siccar Point journey was memorably described by Playfair (1805, 71–73). The three men knew that if the rocks were interpreted according to Hutton's theory, a vast amount of time was entailed. The sediments of the schistus/greywacke must have been deposited horizontally under the sea and consolidated by heat and the pressure of superincumbent material. Then the area was upheaved by forces acting from within the Earth, being hardened by heat and pressure in the process. The elevatory forces—or perhaps ones that acted subsequently—were so immense that the strata were made to stand almost vertically. But lifted up thus, they were then subjected to weathering and erosion, and the upheaved rocks were eventually reduced to a more or less level surface. In time, the (now vertical) strata subsided below the sea (by a process that was not adequately explained) and were covered by layers of sediment transported from some adjacent high ground. Again there was consolidation, following which uplift occurred, but without

Figure 5.6: Unconformity at Siccar Point. Nearly horizontal sandstones overlying steeply inclined greywackes. British Geological Survey, photoarchive P218993. IPR/63–35C British Geological Survey © NERC. All rights reserved.

the overlying sandstones being bent or inclined. Finally, weathering and erosion (including in this case the action of the sea) operated on the rocks to produce the exposure that the "philosophers" observed. And thus the history of the rocks now observed at Siccar Point could be understood—provided one could draw on virtually unlimited drafts of time. As Playfair (1805, 73) eloquently wrote,

> Revolutions still more remote appeared in the distance of this extraordinary perspective. The mind seemed to grow giddy by looking so far into the *abyss of time;* and while we listened with earnestness and admiration to the philosopher who was now unfolding to us the order and series of these wonderful events, we became sensible how much further reason may sometimes go than imagination can venture to follow. [italics added]

Clearly, cyclic thinking was at work here. Hutton provided evidence in favor (but not formal proof) of the Earth's great age and the cyclic nature of geological processes.

That day in 1788 was one of the most important in the history of geoscience, and Siccar Point has long been recognized as one of geology's most significant sites. The day was especially important in that Hutton gained Playfair as a convert to his way of thinking, and it was Playfair (1802) who successfully popularized Hutton's ideas, whereas Hutton's prolix style, his endless quotations in French, and his confusing theory of heat (see the discussion later in this chapter) did not gain him many adherents. Needless to say, the rocks at Siccar Point would have been observed many times before Hutton and his friends examined them. But the rocks signified an "abyss of time" only when viewed through the lens of Hutton's theory.

Regarding a theory of heat, in his old age Hutton endeavored to give some kind of physicochemical explanation of the forces causing elevation and subsidence, dependent on his understanding of the nature of heat. But he had rather little success. He knew that bodies expanded when heated, and the heat that produced this effect he called "sensible heat." He also knew, from Joseph Black's work, that when heat was applied to a solid, that solid might increase in temperature, but on reaching its melting point, it would then melt *without* change of temperature even though still being supplied with heat. In such a "change of state," the heat supplied to produce melting was called hidden heat, or "latent heat." (Latent heat was also involved when a liquid was vaporized by heating.)

But the *nature* of heat was uncertain. We regard it as a form of energy, but Hutton thought it was a kind of weightless "substance" (called "caloric" by some people though not by Hutton). He knew that ordinary inertial objects have mass and that massive bodies are attracted to one another by gravitation (as Newton had shown). But there also seemed to be repulsive forces at work in the universe, as when water is boiled, for example: Watt's engines exerted *pressure* in their cylinders.

We also distinguish between radiant heat and heat that is transmitted through a body by diffusion (or conduction). Hutton had no adequate concept of radiation, but he knew that the Sun's heat shines across space. He called it "solar substance," which, though weightless, was absorbed by plants, though he did not know how. (Plants grew heavier as the Sun shone on them, as he knew from his agricultural experience, though water and soil were necessary, too; but obviously Hutton didn't know how photosynthesis worked.) Adding to the complications, Hutton accepted the old "phlogiston theory" of combustion (which was collapsing at the end of the eighteenth century), according to which an inflammable material contains a weightless "substance" or "principle" called "phlogiston," which is dispersed into the atmosphere during combustion, the air supposedly being necessary to absorb the emitted phlogiston. Hutton was inclined to suppose that "solar substance" and "phlogiston" were one and the same. But others thought that phlogiston might be "inflammable air" (our hydrogen), or hydrogen could be "water + phlogiston." It was all a muddle!

Hutton grappled with such issues in two books: *Dissertations on Different Subjects in Natural Philosophy* (1792) and *A Dissertation upon the Philosophy of Light, Heat, and Fire* (1794a). We cannot follow the details here, but he tried out the idea that objects normally *attracted* one another according to the inverse square law of gravitation ($F \propto 1/d^2$). Thus, he spoke of "gravitating *matter*." At very close quarters, however, objects supposedly began to *repel* one another, according to a force law in which the power was greater than two ($F \propto 1/d^{>2}$; see Hutton (1792, 627–628). The repulsive force (or solar substance!) could take various guises: "sensible" heat, manifested by expansion, latent heat, light, electricity, or phlogiston. So when sediments were under extreme pressure, they might turn from contraction under compression to expansion. That is, in the geostrophic cycle there could be alternating periods of contraction (compression or consolidation) and expansion (producing land elevation) due to differing balances of forces (or a predominance of different hypothetical weightless "substances") (Gerstner 1968).

All this was speculative, and is difficult to understand. Hutton lacked the concept of energy, so the story that I have tried to describe above hardly makes sense to us (though if one regards phlogiston as energy, some of the problems that Hutton was trying to understand fall into place). So we see that, despite all his successes in seeing into the "abyss of time" and his successful predictions, his theory had problems. The physical explanation of expansion and uplift was not presented in his *Theory of the Earth,* and it attracted little or no following. Expansion was the Achilles' heel of Hutton's theory, and the problem remained unresolved for generations. In practice, expansion and uplift were assumed by Huttonian cyclists largely on the basis of field evidence rather than in terms of physical theory.

So, regarding his theory of heat, Hutton's contemporaries found it difficult to understand what he was driving at. One thing was certain: he did not think that the Earth's internal heat was due to combustion of materials within the Earth. Were this the case, a central fire would eventually produce an equilibrium temperature distribution in the Earth rather than act intermittently or with episodic changes (as Hutton's theory seemed to require) and also (presumably) global expansion. In addition, it might be expected to run out of fuel eventually, so there *would* be a "prospect of an end"!

As Gerstner (1971) has emphasized, Hutton's model depended on a balance of attractive (gravitational, cohesive, and concretive) and repulsive forces (specific [or "sensible"] and latent heats). (All this was different from the caloric theory, popular at the time, though Hutton's "solar substance" was somewhat akin to the hypothetical "matter of heat" or "caloric.") There could be different states arising from the opposed forces producing elevation (expansion) or subsidence (contraction) at different times and places. But when Hutton started talking about "solar substance" in reference to solar radiation (as we would say) but thought that this "substance" was immaterial, confusion and misunderstanding were likely, despite the quite widely accepted notion of "caloric." So people could accept the geostrophic cycle while discounting the physical basis of Hutton's theory. The general acceptance of Hutton's theory took time. Geological theory was racked with controversy until well into the 1820s. As the professor of natural history at Edinburgh—Robert Jameson (1774–1854), a student and disciple of Werner—gained control of geology teaching at Edinburgh (and even of Hutton's specimens), Huttonian theory tended to be eclipsed in Scotland for decades despite the efforts of Playfair and Hall. (But it was Jameson who introduced the term "unconformity" into geological vocabulary.)

From the foregoing account, it would appear that Hutton worked out his cyclic theory essentially on his own, but it is likely that he knew of Hooke's theory of pole wandering, even though he never acknowledged it directly. Drake (1996, 120–28) has drawn attention to the fact that Hutton (1788, 222–223) tried to show that Hooke's pole wandering theory was unsatisfactory:

> [L]et us suppose the axis of the Earth to be changed from the present poles, and placed in the equinoctial line [which would be in the plane of the ecliptic], the consequence of this might, indeed, be the formation of a continent of land about each new pole, from whence the sea would run towards the new equator; but all the rest of the globe would remain an ocean. Some new points might be discovered [exposed above sea level], and others, which before appeared above the surface of the sea would be sunk by the rising of the water; but on the whole, land could only be gained substantially at the poles. Such a supposition as this, if applied to the present state of things, would be destitute of every support, as being incapable of explaining what appears.
>
> But even allowing that, by the changed axis of the Earth, or any operation of the globe, as a planetary body revolving in the solar system, great continents of land could have been erected from the place of their formation, the bottom of the sea, and placed in a higher elevation, compared with the surface of that water, yet such a continent as this could not have continued stationary for many thousand years; nor could a continent of this kind have presented to us, everywhere within its body, masses of consolidated marble, and other mineral substances, in a state as different as possible from that in which they were, when originally collected together in the sea.

Thus, Hutton evidently knew about Hooke's theory, though he chose not to name his predecessor. Drake (1996, 127) suggests that Hutton was "totally confused" about Hooke's argument. I confess I do not read him thus. Hooke thought he could account for land being sometimes above and sometimes below sea level, but his work failed to explain why limestone might be converted to marble, which for Hutton evidenced the agencies of heat and pressure. Drake has suggested that Hutton dragged Hooke's theory into his account gratuitously since he seemed disposed only to reject it. However, one could also say that Hutton was mentioning an earlier theory, probably known to his auditors, which he wished to dispose of before expounding his own heat-based theory. (But if that were the case, Hutton should have cited Hooke's theory directly.) It remains uncertain, then, whether the cyclic aspects of Hutton's theory were indebted to Hooke. On the whole, I'm inclined to think they were not (or not much). His medical theory, the steam engine, and his knowledge of erosive and depositional processes at work in Berwickshire and on its coastline and elsewhere provided, I think, the basis of his theorizing.

Later in this book (Chapter 13), we shall refer to the so-called Gaia hypothesis, according to which the Earth may be regarded as a self-sustaining, "quasi-living" system. Some commentators on, or adherents of, the Gaia hypothesis have seen Hutton as an early exponent of such a view (e.g., Westbroek 1992, 212). For the Gaia theorists, living organisms are essential to the whole cyclic process, operating within, on, and around the

Earth, with the oceans, the atmosphere, and living organisms all involved. Hutton's views did not, however, emphasize the role of life. His system was more mechanical and focused on the Earth's internal heat. But he did know about the role of living organisms in breaking down rocks into soil, and his consideration of "solar substance" demonstrates his recognition of the importance of plants in capturing the Sun's energy (to use an anachronistic term), making life on Earth possible as a self-sustaining system.

Here, in the same chapter as Hutton, I should like to consider also some ideas of the French "philosopher of evolution" Jean Baptiste Lamarck (1744–1829). The two philosophers are not often thought of in the same breath, but their ideas had features in common. Their dates overlap. Both were deists. Both had evolutionary or "transformist" ideas—Lamarck's mechanism for evolutionary change having to do primarily with changes made by organisms as changes in response to changing environments that could be passed on to later generations ("inheritance of acquired characteristics"), whereas farmer Hutton knew that animals could be changed from one generation to the next by artificial selection or selective breeding, and, like Darwin, he thought that something similar might occur naturally with the best-adapted animals tending to survive and propagate themselves. (This idea was expressed in an unpublished manuscript, "Principles of Agriculture," which survives in Edinburgh.)

Both Hutton and Lamarck envisaged the Earth as being of almost unlimited age, and their theories of the Earth were both cyclic, though Lamarck emphasized the role of water, not heat. In addition, both had thoughts about pole wandering, and chemical considerations played a significant role in their geological theorizing, though in different ways.

Though he traveled extensively in his earlier years, at the time of his main geological and biological theorizing, Lamarck was essentially a museum man: from the time he was 50, he held a professorship at the *Muséum d'histoire naturelle* in Paris, with responsibility for the study of invertebrates and the identification and classification of fossil shells. There was discussion in Paris at the end of the eighteenth century as to whether these might have modern counterparts living in unexplored parts of the oceans. The fossils were obviously different from modern forms, but Lamarck could not think of a mechanism whereby a species could become extinct, as there was (so it was thought in the eighteenth century) always a "balance" in nature. So transmutation, or gradual evolution, *must* have occurred if the fossil evidence was to be explained. This unpopular idea was expounded in his *Recherches sur l'organisation des corps vivans* (1802), *Philosophie zoologique* (1809), and *Histoire naturelle des animaux sans vertèbres* (1815–1822). There is no evidence of direct contacts between Hutton and Lamarck.

For present purposes, however, it was Lamarck's strange work *Hydrogéologie* (Lamarck Year 10/1802, 1964) that is of major concern. The book is inconsistent, but according to its English translator, Albert Carozzi, Lamarck sent off the four chapters separately to be printed but changed his mind on some issues in the time between starting and finishing the book—hence the inconsistencies!

Like Hutton, Lamarck accepted that weathering and erosion were constantly eating into the landscape and that sediments were deposited in layers in the seas. However, he imagined that apart from the initial ocean basins, the Earth's surface was originally an immense flat plain but that hollows formed on its surface because of the action of rain and running water. Mountain ranges were formed by the hollowing out of adjacent river basins. Additionally, some mountains originated as volcanoes. Dipping strata were thought to be the result of sediments being deposited on the inclined sides of marine basins, or they could have been produced by subsidence.

Lamarck thought that movements of the oceans were caused by the actions of the Sun and the Moon. The Moon produced the tides, but the position of the maximum tidal "bulge" of the Earth's envelope of water changed through the month. The scouring movements of the water acting on the ocean basins tended to deepen them and also move the sediment around in the basins. Moreover, because the oceans do not form a universal liquid envelope around the Earth, the geometrical center of the Earth and its waters does not coincide exactly with its center of gravity. (This sounds like Buridan again.) So Lamarck imagined that the Earth's center of gravity made a slow circuit around its geometrical center, for the ocean basins were making a slow westward circuit around the globe, redistributing sediment during their progression. The Gulf of Mexico, for example, supposedly evidenced the slow westward movement of the Atlantic Ocean, whereas the low-lying parts of western France represented an area fairly recently abandoned by the Atlantic. Lamarck's theory accounted for the presence of inland fossils, if nothing else!

Further, since the westward movement of water was blocked by the Earth's two largest landmasses, this would mean that water would tend to be driven southward, so that there are unusually strong currents near Cape Horn and the Cape of Good Hope, and there is a predominance of oceans in the southern hemisphere (at least at present). So, suggested Lamarck, the changes of the balance of waters and sediments and the changes in position of the center of gravity would induce pole wandering, so regions could experience different climates at different epochs (evidenced by the fossils discovered).

Toward the end of his *Hydrogéologie*, Lamarck developed further the idea of the possible mechanisms for mountain formation arising from pole wandering. The Earth has an equatorial bulge. So, if poles "wander," this bulge will be displaced and may rise above the oceans to form mountain ranges. In fact, he began to contemplate complete revolutions of the pole points around the surface of the globe (which is reminiscent of Hooke, though Lamarck didn't mention him). The Earth was, Lamarck suggested, partly plastic, so that it could modify its shape to match the form required according to the situation of its poles. But if (as Lamarck supposed) the North Pole were moving toward Europe, one could equally say that Europe was moving northward. So the great east–west mountain ranges of Eurasia could be the remains of a former equatorial bulge! Such ranges would not appear further north, as they would have been obliterated by the plastic Earth adjusting itself to the different "effective" gravitational field at the higher latitudes while a new bulge is forming in the present equatorial regions. So again we have a cyclic theory.

But there was more to it all than what I have just outlined because, for Lamarck, living organisms played an essential role in the whole process (and it is for this reason that he too is admired by Gaia aficionados, though his theory may appear weird to most modern readers).

Just as Lamarck believed that there was an evolutionary continuum between different organisms, so also he envisaged a continuum between mineral substances. And mineral types could "evolve" one into another by the gradual chemical breakdown and simplification of the residues of living organisms. The scheme is shown in Figure 5.7. Lamarck rejected the new antiphlogistic chemistry of Antoine Lavoisier and his coworkers, preferring the Aristotelian idea of four interchangeable "elements" (earth, water, air, and fire). So he could entertain the transmutation of one mineral kind into another.

Thus, granite was not a "primitive" or primeval rock for Lamarck. He supposed that its chief constituents were quartz, feldspar, mica, and tourmaline. The constituent parts of these were liberated into rivers by their breakdown by living organisms and the weathering of rocks, and the parts supposedly came together to form crystals of quartz, feldspar, and so on. Thus, granite precipitated out of aqueous solution, and if the materials were deposited in layers, the product would be gneiss! So as a shoreline receded, thought Lamarck, a river delta would be extended seaward, and a linear deposit of granite would be formed along the river's course toward the sea. Later, the granite deposits would form submarine chains of granitic mountains, and (after appropriate pole wandering) the granite

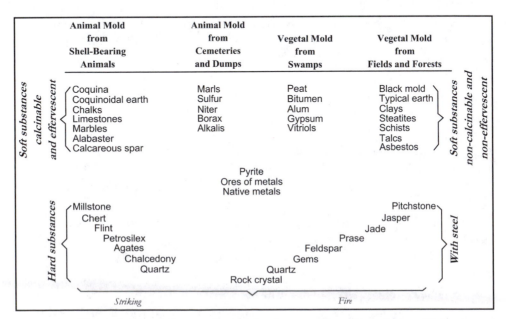

Figure 5.7: The evolving continuum of mineral substances from the residues of living organisms, according to Lamarck's *Hydrogeology* (1964, 119). Reproduced by courtesy of Albert Carozzi. Redrawn from the original by Ricochet Productions.

might be exposed as the cores of mountain ranges and then weathered and eroded, as in Hutton's theory.

One can hardly find a more fanciful theory in the history of geology, and Lamarck's geo-theorizing was forgotten or decried for many generations. But it had one key thing that was right: the recognition of the role of life in geological processes on Earth, which makes our geology so different from that of the other planets. This issue will be explored further in Chapter 13.

If we look for analogies with the three fables referred to in Chapter 1, I suppose Lamarck's theory comes closest to the sad story of "Ilkley Moor b'aht 'at." Organic matter is constantly cycled. But while, for Lamarck, one specimen of granite might be much the same as another, this was not so for living organisms. They seemingly evolved toward higher states of "perfection." While Lamarck's mineral cycle seemingly just went around and around, his organic cycle had a definite "forward" or "progressive" trajectory, and it is this aspect of his thought that is chiefly remembered today.

NOTE

1. According to Werner, the primeval Earth had an irregular core covered by a universal ocean, the dissolved materials of which were successively deposited on the core, high and low. The first precipitate was granite. As the water level decreased, the higher areas were exposed, covered by granitic matter. Weathering and erosion of the granite provided sediment for the deposition on sloping surfaces of laminated rocks such as schists—thought to be a combination of sediment and additional precipitated materials. These, along with other sediments, formed Werner's "Transition Series" (*Uebergangsgebirge:* in Germany mostly Carboniferous–Jurassic in modern parlance). After further lowering of sea level and exhaustion of most of its dissolved chemicals, the subsequent deposits were noncrystalline, mostly horizontal, sediments. After the water subsided still further, the present seas are remnants of the initial ocean. Basaltic dykes and other "vein" rocks were supposedly deposited *from above* in cracks or rents formed in strata as they slipped on inclined surfaces. Werner's theory was generally "directionalist" though somewhat more complicated than just described. The ocean level generally fell but rose at times (for reasons that were unclear) and then deposited crystalline basaltic material on hilltops or in the rents in the slumped beds. For details, see Werner (1971). The theory sometimes matched observations, where mountains have granitic cores flanked by inclined schists and then by more or less horizontal sediments. Werner could have given an explanation of the structure of Arran (Figure 5.2).

6

CHARLES LYELL: AN EARTH ALWAYS CHANGING BUT EVER THE SAME

By the early years of the nineteenth century, geology was a recognized member of the family of natural science, being concerned with elucidating the Earth's history (Rudwick 2005), and people who studied geology were called geologists. Charles Lyell (1797–1785) was (in the opinion of Anglophones!) the most important geologist of the nineteenth century, and his *Principles of Geology* (1st ed. 1830–1833, 11th ed. 1872) exerted a major influence on geology and shaped Darwin's thinking.

Lyell was born into a well-to-do family in Forfarshire, Scotland, but much of his youth was spent at the family's second home in Hampshire. He studied mathematics and classics at Oxford, where he became interested in geology through the lectures of William Buckland (1784–1856). Lyell began making geological observations around his family's Scottish estate, and during his vacations he traveled with his family on the Continent. Even while a student, he was elected to the Linnean Society and the Geological Society of London (founded 1807).

After graduation, Lyell worked in a law office in London, but, being less interested in law than science and having independent means, he began to relinquish his legal work and became a "gentleman–geologist." He was elected to the Royal Society in 1826 and in 1828 ceased legal work to travel on the Continent, meeting such figures as the comparative anatomist Georges Cuvier (1769–1832) and the conchologist Gérard Deshayes (1797–1875) in Paris. Subsequently, Lyell acquired significant income from his geological writings. In 1832, he married Mary Horner, daughter of Leonard Horner, himself a geologist, who had learned Huttonian theory in Edinburgh. The couple settled in London, where Lyell became one of its leading scientists. He was appointed professor at King's College, London, and gave lectures there in 1832–1833 but didn't continue, not enjoying teaching and having financial independence.

Buckland taught Lyell the essentials of stratigraphy and particularly the idea that strata could be identified and correlated by their fossil contents (a principle enunciated by the William Smith [1769–1839][1].) In the religious atmosphere of Oxford and trying to show that his science was compatible with the Bible, Buckland specialized in the study of superficial deposits and cave remains, and suggested how such materials could be explained by the Noachian Flood, which was thought to have occurred only a few thousand years ago. This position became known as "diluvialism," for the blanket of deposits that cover much of northern Europe was referred to as "diluvium," being thought to be the product of the Flood—not debris of glacial origin as supposed today. Buckland distinguished between "alluvium" (sediment deposited by normal river floods) and "diluvium"—the material supposedly deposited by a "one-off" catastrophic flood. But such a global flood would have been impossible according to the laws of nature as presently acting and incompatible with the geological processes seen at work today. Lyell came to regard Buckland's theory as unsatisfactory, and Lyell's geological career became an extended effort to counter such ideas (though he was a Christian).

Lyell traveled in Scotland in the 1820s, visiting such sites as Glen Tilt and Siccar Point, and began to incline toward the geological thinking of Hutton, perhaps under the influence of his father-in-law-to-be. Werner's theories were rejected as being incompatible with the known insolubility of most rocks in water and with Hutton's field observations. In Forfarshire (which he mapped in 1824), Lyell saw marls being or already deposited in freshwater lakes fed by springs and associated with shells and plant remains. He knew that similar freshwater limestones occurred in the Paris region, which French geologists thought had no modern analogues. Lyell's observations showed that this was not the case, and so his mind turned toward the principle of explaining phenomena in terms of presently occurring processes as the proper method for geology.

As a Huttonian, Lyell had to consider the question of geological time. His arguments for the Earth's antiquity were developed during his journey in 1828 to Sicily, where he studied the huge volcano Etna, which had seemingly accumulated from successive lava flows (Rudwick 1969). Information about recent flows gave an idea of their rate of accumulation and the mountain's growth rate. So, knowing its height, one could estimate its age. Further, Lyell examined shells in *recent-looking* strata, which could be seen cropping out *below* Etna's lavas. They nearly all occurred in today's Mediterranean. So if geologically recent rocks were ancient in human terms, rocks lower in the stratigraphic column must be *exceedingly* ancient. Hence, the Earth was enormously old. In this argument, Lyell assumed that Etna's flows occurred at a rate approximately equal to present flows. In saying this, he was *assuming* that nature's past operations were analogous to those operating in the present. This guiding principle was later dubbed the principle of uniformity. Lyell was a "uniformitarian" with regard to Etna, as with respect to the lake deposits in

Forfarshire. He adopted "uniformitarianism" as a general rule for geological theorizing.

Lyell did not publish his ideas about Etna's age in the 1830s, but in a letter to his sister written in 1830, he suggested that the shells found in sandy limestones cropping out below the lavas were perhaps 100,000 years old, and he expressed similar ideas in his King's College lectures (Tasch 1975, 1977). Years later, with the advent of the astronomical theory of glaciation (see Chapter 9), Lyell sought to estimate the time since the Glacial Period (Pleistocene) and obtained figures in reasonable agreement with modern values (Tasch 1977).

Also during his Italian journey of 1828, Lyell visited Pozzuoli by the coast near Naples. There he observed three *standing* columns of a Roman building. They had borings of marine organisms the same distance up the columns, implying that the land there had fallen below sea level since Roman times and had subsequently risen, and this had happened without the columns toppling. (Alternatively, the sea level might have risen and fallen, but this seemed less likely, as similar evidence was not known elsewhere around the Mediterranean.) Hence, Lyell inferred that land could rise or fall at different places, much as Hutton had previously proposed (though Hutton didn't say much about subsidence). Moreover, the changes at Pozzuoli had been gradual, for, as said, the columns had not fallen over. So, using the terminology of Rudwick (1971), Lyell was a "gradualist" as well as a "uniformitarian."

On his return to Britain, Lyell began to write a treatise that sought to establish the "correct" practices for geology and enunciate its fundamental theoretical principles— hence its title: *Principles of Geology*. These may be summed up by the adage (as later stated by Archibald Geikie [1962, 299]) that "the present is the key to the past." In fact, Lyell's book was subtitled *An Attempt to Explain the Former Changes of the Earth's Surface, by Reference to Causes Now in Operation.* The

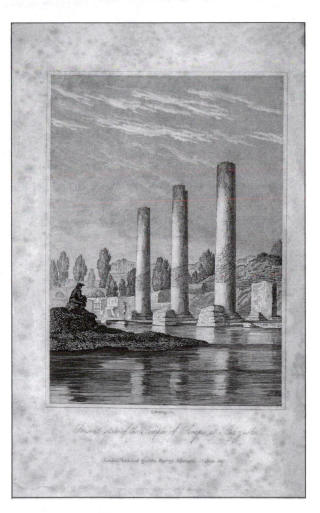

Figure 6.1: The "Temple" of Serapis at Pozzuoli near Naples. Frontispiece to volume 1 of Lyell's *Principles of Geology* (1830). (Today it is not thought to have been a temple.) Image copyright History of Science Collections, University of Oklahoma Libraries.

Pozzuoli observations were so important to Lyell that he used an engraving of the columns as the frontispiece to volume 1 of his *Principles* (Figure 6.1).

As we have seen, Hutton's geology envisaged grand cycles of rock formation, erosion, transport, deposition, consolidation, and subsequent elevation by intrusion of molten rock from below. The rocks of each cycle and their changes were not necessarily identical in any given place, and the geologist needed to work out the history of what had happened at each locality. But Hutton made little attempt to do this, and overall Hutton's Earth did not have an historical direction: it did not "progress." Although different in their details, things were much the same in the past and present (humans excepted).

Lyell's views were similar, but he placed greater emphasis on fossils and supposed that conditions were constantly changing at any given locality from one epoch to another because of local changes of relative levels of land and sea. Climate changed concomitantly, according to whether more or less high land happened to be nearer the poles or the equator at any given time, the former state of affairs producing lower global temperatures. Particular species disappeared if they failed to meet the changing conditions.

But to maintain a steady-state situation, new species needed to come into existence from time to time, replacing those that had become extinct. Lyell presumed that this occurred, though he did not know how, nor had anyone observed the formation of new species (thus contradicting Lyell's own major "principle"). But he rejected Lamarck's suggestion that one species could "transform" or "evolve" into another. Further, Lyell assumed that the basic animal types (mammals, reptiles, and so on) had always been present. On this view, there was a gradual turnover of species through time. One can compare his model with the population of a school. New pupils arrive each year, and old ones depart. The overall number may be approximately constant, so in a sense the population remains the same. Nevertheless, the school has a history. Even if its population is approximately constant, its performance may change with the changing socioeconomic status of its catchment area, the arrival of migrants, its funding, the construction of new buildings, fires, new teachers, or whatever. (I'm thinking of my own school!)

Lyell's model for the coming and going of species is represented in Figure 6.2. His Earth could have a stratigraphic *history*, which could

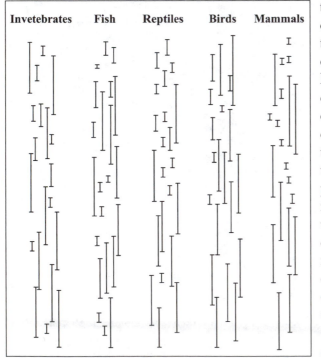

Figure 6.2: Illustration of Lyell's theory of "random" creation and extinction of species.

be investigated by the geologist. Lyell was greatly interested in human history, and this interest spread into his geohistorical work (Rudwick 1977), making his geology significantly different from Hutton's, even though it was its intellectual descendant.

According to this model, Lyell assumed that the further back the geologist traveled in time, the fewer extant species would be found. In fact, nearly all species before the beginning of the Tertiary in Europe were now extinct. Furthermore, the Tertiary could be subdivided according to its proportions of extant fossils, in what Rudwick (1978) has called a "statistical paleontology." The subdivisions of the Tertiary strata that Lyell proposed in volume 3 of his *Principles* (1833) were as follows:

[Quaternary: approximately equivalent to the Pleistocene]
[Tertiary]

Newer Pliocene	96% recent fossils
Older Pliocene	49% recent fossils
Miocene	18% recent fossils
Eocene	3% recent fossils

[Secondary: approximately equivalent to the Mesozoic]
[Primary: approximately equivalent to the Paleozoic]

The division of strata into Primary, Secondary, and Tertiary had been suggested by the Italian geologist Giovanni Arduino in 1760. The Frenchman Jules Desnoyers (1801–1887) suggested the name "Quaternary" in 1829, and Lyell himself proposed the term "Pleistocene" in 1839. The terms "Mesozoic" and "Pal(a)eozoic" were suggested by the English geologist John Phillips in 1840 and 1841, respectively.

Between the Secondary rocks and the Tertiary (the original subdivisions of the latter are in the previous list), there was seemingly a period of nondeposition in Europe, so that all the organisms represented by Secondary fossils were now virtually extinct. There had been a complete turnover of forms in Europe during the stratigraphic time gap between the Secondary and Tertiary strata. (But deposition had been occurring elsewhere in the world during the time of the stratigraphic hiatus in Europe.) Likewise, there was thought to be a large time gap and turnover of forms between the Primary and the Secondary eras.

As can be seen from Lyell's percentages, the further back in time, the smaller was the proportion of forms resembling those still living today (likewise from top to bottom of both the Secondary and the Primary strata). But, as indicated in Figure 6.2, Lyell thought that the main animal types had always been present. If mammals were absent from the early strata, this was due to their loss by erosion or the poor preservation of terrestrial animals. Even so, some small marsupial remains had been found by a quarryman in 1812, in Secondary (Mesozoic) rocks near Oxford, and were examined by Buckland. Lyell thought that such evidence supported his idea that the major animal types extended into the indefinite past.

It is interesting that Lyell thought that the kinds of species produced at any given time were related to the environmental conditions. For example, if particular past climatic conditions returned, so too might now extinct animals—such as the giant iguanodon (first studied scientifically by Gideon Mantell in 1820):

> The huge iguanodon might reappear in the woods, and the ichthyosaur in the sea, while the pterodactyle might flit again through the umbrageous [shady] groves of tree ferns. Coral reefs might be prolonged beyond the arctic circle, where the whale and narwhal now abound. Turtles might deposit their eggs in the sand of the sea beach, where now the walrus sleeps, and where the seal is drifted on the ice-floe. (Lyell 1830, 123)

We see from this interesting quotation that Lyell's *historical* Earth had *no overall direction.* (It could reverse itself!) His geology was "nondirectionalist."

Lyell's ideas attracted much attention, though most contemporary geologists couldn't accept that there was no progress in the fossil record through time (Bartholomew 1976). On the other hand, his desire for geology to have its own principles, with geological processes operating in accordance with the present laws of nature, found approval, as did his mastery of stratigraphic detail. Lyell wanted to ensure that geology was a science in its own right, distinct from cosmology or cosmogony. Geologists, he thought, didn't need a general "theory of the Earth" such as his eighteenth-century predecessors had sought to offer.

Lyell was much interested in French as well as Italian geology. He acknowledged Cuvier's mastery of paleontology and comparative anatomy (see pp. 59 and 212) but rejected his "catastrophist" (and "directionalist") theory, which envisaged a series of "catastrophes" of uncertain cause that had, through the geological record, wiped out many or most species. The world was then repopulated (again by a process unknown) by new species. Cuvier's theory was propounded in order to account for the sudden discontinuities evident in the paleontological/stratigraphic record (the last of which had encouraged Buckland to believe that the Noachian Flood was a major geological event). Lyell's *Principles* did much to counter such doctrines, especially in Britain. (On the importance of Cuvier in the development of stratigraphy and geology more generally, see Rudwick [2005].)

Lyell had met the conchologist/paleontologist Gérard Deshayes in Paris in 1828 and was assisted in the identification and stratigraphic placement of the shells that he collected that year. Deshayes also supplied data from other parts of the world that provided evidence for Lyell's theory of the gradual turnover of forms. On the other hand, Lyell reacted negatively to the tectonic theory of Léonce Élie de Beaumont, which envisaged mountain ranges as having been formed as a result of sudden crumplings of the Earth's crust as it cooled and contracted (see Chapter 8). Lyell significantly hindered the acceptance of this French doctrine in Britain, though some geologists such as Adam Sedgwick, professor of geology at Cambridge, and Henry De la Beche, director of the Geological Survey of Great Britain, were attracted to the idea and tried to apply it in their work. But Lyell objected to thinking of a geological "formation" as a time-bounded tectonic (mountain building) event rather than a period of accumulation of strata with characteristic fossils. Lyell's usage has prevailed regarding the meaning of "formation."

In volume 2 of *Principles,* Lyell attended closely to Lamarck's "transformist" (evolutionary) theory. According to Lamarck, changing conditions produce new needs for organisms. To adjust to changing circumstances, organisms may alter their habits and consequently their forms. Furthermore, these changes may, according to Lamarck, be inherited by subsequent generations, producing a gradual transformation of species. He thought that the first simple forms of life appeared by spontaneous generation, without divine action.

Lamarck's ideas were expounded and then rejected through many pages of Lyell's *Principles.* He objected that the stratigraphic record did not reveal smooth transitions, such as Lamarck's theory would lead one to expect. New species could not be produced by breeding (though new varieties could). Hybrids were sterile. (But note that Lyell's thoughts on the iguanodon's return were somewhat Lamarckian in character.) Lyell devoted much effort to thinking about the concept of species, the possible "laws" of their distribution, and the extent to which they could or could not show modification due to environmental changes. The problem of species and speciation was one of the main features of his book, and it set the scene for Darwin's work on the "the origin of species."

A major problem for nineteenth-century geologists was the large quantities of superficial deposits: gravel, tenacious clay containing unsorted rock fragments and fossil remains, and large boulders of rock distant from the nearest "solid" outcrops of rock of that type (so-called erratics). As mentioned, Buckland ascribed the deposition of such materials to the "Deluge." Such "diluvial" phenomena were explained by the work of the Swiss geologist Louis Agassiz (1807–1873) and his theory of an ice age in the late 1830s and early 1840s (see also Chapter 9).

According to Agassiz, there had formerly been a colder climate with northern Europe covered by ice, which had transported boulders, ground up the underlying rock, and deposited it, along with river gravels, over the land. The ice also could have scratched the underlying rock and transported shells to hilltops. Agassiz lectured on these matters in Glasgow in 1840, and some geologists, including Lyell, were converted to his ideas.

As said, Lyell's theory envisaged periods of cold if much high land happened to be situated in the polar regions; so the idea of an "ice age" was not unpalatable to him. In fact, on the very page where he had written about the iguanodon, Lyell had also referred to the "great year." That is, he envisaged great cycles of climatic change, tracking the random changes of the disposition of high land over time.

But Lyell's conversion to Agassiz's doctrine was short lived. The theory seemed to take one too far from present analogies or present climatic conditions— involving departure from the general uniformity of conditions over time.

So Lyell adopted a theory that came to be called "glacial submergence"—there had been a period of great cold but not such as to produce an all-enveloping mass of land ice. Rather, there was a general fall of land surface, causing marine submergence, accompanied by cooling, causing extension of ice fields and the transport of boulders by drifting icebergs (hence the diluvial deposits were, and still are, generally called "drift"). During a visit to North American in 1845, Lyell saw floating ice in the St. Lawrence River—which modern observation seemed to support the occurrence of erratics satisfactorily, in accordance with his methodology. He did eventually accept Agassiz's ideas for Alpine regions but not the general land-ice theory.

After Darwin returned from his *Beagle* voyage in 1836, he and Lyell became close friends, but before the publication of *The Origin of Species* (1859), Darwin mostly kept his emerging transformist ideas to himself. Lyell was opposed to transformism for reasons that he developed back in the 1820s and 1830s, and like many people, he was concerned about evolution's supposed implications for religion and hence for social stability. In his Presidential Address to the Geological Society (1851), he spoke against evolutionary ideas. "Man," he thought, was a recent creation, subsequent to the mammoths. However, after Darwin revealed his ideas to Lyell about 1856, he was reluctantly converted and assisted the first publication of Darwin's ideas in 1858. This entailed Lyell giving up his long-held "nonprogressionism." In any case, the evidence against it was growing, especially with the failure to find mammals and birds in Paleozoic rocks. So in his *Geological Evidences of the Antiquity of Man* (1863), Lyell set forth ideas on transformism and stated his acceptance of Darwin's theory of evolution by natural selection, though representing it as a "modification" of Lamarck's doctrine. Lyell's program for the study of the cycles of life was broken at last. It entailed an intellectual wrench for him, as the "catastrophists" against whom he had labored for many long years were also "progressionists" (with "advances" made at each creation of forms after a "catastrophe"). Darwin's theory at least did not invite him to accept mysterious new creations, and up to a point the present could still be thought of as the key to the past, though the emergence of new species was not observed directly in the nineteenth century.

On reading Lyell, one is struck by his command of the literature, especially in stratigraphy. His influence was great, both in his own day and subsequently. There was (and is), however, ambiguity in the concept of "uniformitarianism" (which can encompass gradualism, "steady-stateism," naturalism, and "actualism"—or the idea that *actually* observable processes should be used in geological explanations). Lyell held to all these positions. Modern pedagogues commonly refer to uniformitarianism without making the foregoing distinctions. And modern geology does not necessarily adhere to any of them, apart from naturalism, except in its rhetoric. But Lyell convinced people that *his* approach was the right one, if geology was to be a science.

We can see cyclic ideas at work in Lyell's thinking: in his adoption of Hutton's "geostrophic cycle," the idea of a slow turnover of organic forms, and associated with this the "great year" of global climate changes. Lyell's arguments for the Earth's great age, drawn from observations of a geologically recent volcano, were of highest importance. Geologists could not do without time—and a great deal of it. But after Darwin, the old Lyellian cycle was no more.

Lyell's theory never envisaged an exact repetition of events of the "captain and mate" variety. In fact, he wrote to a correspondent in 1835, "My notion of uniformity in the existing causes of changes always implied that they must for ever produce an endless variety of effects, both in the animate and inanimate world. I expressly contrasted my system with that of 'recurring cycles of similar events'" (Lyell 1881, 2:8–9). So, as Gould (1987, 167) put it, Lyell could be regarded as "the *historian* of time's cycle."

NOTE

1. The surveyor/engineer William Smith (1815) produced the first general geological map of England and Wales, delineating the principal boundaries of the stratigraphic column, each subdivision being represented by a distinctive color. His interests were primarily structural rather than geohistorical, but the map and sections he produced certainly were such as to enable one to "read the historical record of the rocks," at least for southern England.

7

CYCLING IN THE LANDSCAPE

If Huttonian/Lyellian theory is accepted—with land being elevated from time to time and then subjected to weathering and erosion and if, for whatever reason, the elevation is repeated—then this cycle of elevation and land denudation would produce cyclic changes in landforms. This idea was worked out with enthusiasm in a new specialty called geomorphology: the science that treats the general configuration of the Earth's surface and especially landforms and their genetic interpretation. It emerged in the nineteenth century, notably in the United States and especially in the western states, which were explored and studied by a succession of geological explorers during that century. Following this reconnaissance work and the lead of William Morris Davis (1850–1934), geographers and geologists began to think of landscapes as having a natural cycle (birth, maturity, old age), somewhat analogous to the life cycles of living organisms, and a quasi-biological language was used to express ideas about the *evolution* of landscapes. Geomorphological phenomena were viewed through the lens of biological or even Darwinian theory.

Early explorations in the American West were undertaken by Spanish explorers and missionaries, such as Francisco Domínguez and Silvestre Velez de Escalante, in 1776–1777. They were concerned with topography and finding routes in remote and inhospitable country but were not concerned with scientific questions as to how and why landscapes were as they were. American topographic expeditions such as that of John Charles Frémont (1813–1890) followed in the 1840s, and during the second half of the nineteenth century, the expeditions took a geoscientific turn (Şengör 2003). Some of the journeys were associated with finding routes for the construction of railroads.

John S. Newberry (1822–1892) looked at the topography, the geological structure, and the stratigraphy of the West in three expeditions from 1855 to 1859 and was the first to delineate the stratigraphy of the Grand Canyon region.

But the major names from the point of view of geomorphology were Ferdinand Vandaveer Hayden (1828–1887), Clarence King (1842–1901), John Wesley Powell (1834–1902), George Montague Wheeler (1842–1905), Clarence Edward Dutton (1841–1912), Grove Carl Gilbert (1843–1918), and Davis. Hayden was professor at the University of Pennsylvania and later served in government surveys. King worked for the Californian and the Fortieth Parallel surveys and in 1879 became director of the new U.S. Geological Survey. He made the first dinosaur discoveries in the Americas. Wheeler was an army geologist who headed expeditions in the southwestern states in the 1870s, notable for the fine photographs showing the remarkable features of the region. Although he lost an arm in the American Civil War, Powell accomplished a successful passage on the Colorado River through the Grand Canyon in 1869 and again in 1871–1872—single-handed, so to speak! He thought that the topographic evidence suggested that the *course* followed by the river was *older* than the mountains of the surrounding plateau: the Colorado had cut through the strata at about the same rate as that at which uplift occurred. He was appointed director of the U.S. Geological Survey in 1883. Gilbert served with the Ohio Survey and then with Wheeler's expedition west of the 100th meridian. He assisted Powell in the Rocky Mountains and later joined the Federal Survey. He supposed that the "basin-and-range" country of the western states could be attributed to block faulting. Dutton also explored in the Colorado area, later becoming head of the Colorado Division of the Federal Survey. He initiated one of the most important concepts in geology, namely, *isostasy*, which supposes that crustal units of the lithosphere "float" on the underlying, more plastic material, rather like icebergs on the sea. Following earth movements, crustal blocks can achieve a state of equilibrium, "floating" high or low according to their densities. Davis studied engineering at Harvard, was employed at the Argentine Meteorological Observatory, and worked with Raphael Pumpelly on the Northern Pacific Survey. He subsequently returned to Harvard as an instructor and rose to the position of professor of physical geography and then professor of geology. He is sometimes called the "father of physical geography." Davis's whole approach to landscape studies was in terms of their cyclic evolution.

These men observed and recorded the remarkable structures visible in the arid regions of the American West and were convinced of the significance of the importance of rivers in producing erosion and carving landscapes. Powell (1875, 203) introduced the important concept of "base level" for an "imaginary surface inclining in all its parts toward the lower end of the principal stream," below which it is impossible for the stream and its tributaries to cut down further by erosion. The hypothetical limit toward which the erosion of the Earth's surface can progress is, of course, sea level. A base-level surface on land is not normally perfectly horizontal, though it is approximately so in some cases, such as coastal areas. Usually, it is a surface that curves gently upward from sea level. A stream in a mountainous region will carry away all the eroded material that falls into or is picked up by the swiftly flowing water, and erosion will be active on the slope down which the water is running: the stream is cutting down its bed. Further downstream, the amount of sediment and the rate of water flow may be such that sediment is not being deposited and downward erosion is minimal. There is a balance between the amount of sediment arriving and that which is transported away (to form deltas). In such a case, the river or stream is said to be "graded." When an area is eroded down to an almost flat surface and the rivers or streams that flow over it are graded, it is called a "peneplain" (Latin *paene* = almost). The surface of a peneplain has gentle slopes between neighboring rivers, which bear little relationship to the underlying rock structures, as the surface differences have been eliminated by erosion.

However, if a level land surface is, by some means, uplifted—into either a large plateau or a range of steep-sided mountains—then erosion will be renewed, and the rivers will start cutting back from the low-lying ground (perhaps near the sea) into the area of elevated land. Tracing from the sea a river that is cutting back into the landscape like this, one may find a spot where there is a sudden change of slope in its bed, between the old land surface of the plateau on the inland side and the younger land surface, actively eroding, on the seaward side. Such a discontinuity of slope is called a "knick point." It may be marked by a waterfall.

Powell acquired a good sense of such matters from his western journeys. From the way the Colorado River "meandered" through the Colorado Gorge, it was evidently "older" than the plateau country through which it flowed: the river had cut down the canyon at a rate similar to the surrounding rate of uplift. Otherwise, it would not have a winding course through the gorge. The windings were relics of the days when the river flowed across something like a peneplain (the remains of which now form the top of the Colorado Plateau). Many examples of such phenomena can be found. For instance, in New Zealand's North Island, the Manawatu River rises on the sides of the relatively small Puketoi Range to the east of the large northeast–southwest-oriented Tararua–Ruahine Range and cuts right through the mountains at the Manawatu Gorge, reaching the sea at the western coast. So we have an "antecedent" drainage pattern, or an "antecedent" valley, to use the term proposed by Powell (1875, 163). It should be noted that one can glean useful information about the geological history of a region by examining drainage patterns. Thus, the small Puketoi Range must be older than the large Tararua–Ruahine Range.

Powell (1875, 163–66) also introduced the notion of "consequent" streams, or drainage. If a new land profile is established by some means—such as uplift, the formation of a volcano, or whatever—and if new watercourses and drainage system are established on it "in consequence," then, suggested Powell, one can refer to "consequent" streams, or a "consequent" drainage system.

Earlier, in Ireland, Joseph Jukes (1811–1869) had written of "subsequent" streams, valleys, or drainage systems (Jukes 1862). These were ones that developed independently of and subsequent to the laying down of some drainage pattern previously established on the original relief. Subsequent streams might arise, for example, by "river capture," when a stream (A) cuts back so that it meets another one (B) and water from B gets diverted into A. Or A may be more active than B because it is eroding along a fault, or parallel to the orientation of the strata rather than perpendicular thereto. Gilbert (1877, 101) also spoke of "inconsequent drainage," in which streams are not directly related to the underlying rock structures but are, so to speak, "self-guided" or developed on a surface in a random manner. The term was later modified to "insequent" (Davis 1897).

The concept of "grade," mentioned previously, was introduced by Gilbert (1876, 100): "a stream, which has a supply of debris equal to its capacity, tends to build up the gentler slopes of its bed and cut away the steeper. It tends to establish a single, uniform grade." Gilbert also realized that the form of a landscape was dependent on the rainfall. In arid climates, the rate of erosion is limited by the rate of weathering, whereas in humid, well-vegetated conditions, the rate of transport or removal of the weathered material is the limiting factor for erosion. So, as Gilbert (1877, 119), put it, "With great moisture the law of divides is supreme; with aridity the law of structure." That is, in a dry climate, the shape of the land surface is greatly influenced by the hardness of the exposed rock. A humid climate, with the land surface covered by vegetation and soil, will tend to produce a more rounded topography, and the overall drainage pattern, produced by the streams cutting into the slopes of hills and mountainsides, will be shaped according to the lineaments of the mountain ranges.

Gilbert proposed various other geomorphological generalizations. His law of equal declivities stated that when "consequent" streams are established on a hillside of homogeneous rock, they tend to cut channels or valleys for themselves of equal slope angle so that symmetrical profiles of ridges, spurs, and valleys are produced (Gilbert 1877, 141). His law of unequal slopes stated the idea that if one side of a divide has a greater slope than the other, then erosion is more rapid on the side with the greater declivity. So the crests tend to recede from the more actively eroding side, or migrate toward the side where erosion is less active (Gilbert 1877, 140). Thus, for Gilbert, erosive processes were intelligible and exhibited certain *law-like* patterns, so geomorphology could be a "respectable" part of science.

Dutton's work in the American West led him to seek further generalizations about the processes and progress of erosion. Most obviously (at least in the American West), the form of valley profiles depended on the hardness of the rocks that were being carved through by rivers. He accepted Powell's suggestions about antecedent drainages (which Dutton called "persistence of rivers") and base leveling and showed how the erosion of a canyon might proceed according to the hardnesses of the strata (see Figure 7.1).

Dutton always emphasized that stratified rocks were produced from the products of the erosion of other rocks, a principle that he regarded as fundamental to geology. As we have seen, this idea went back as least to Hutton (or Steno), but it had become empirically obvious as a result of the geologically informed explorations of the American West. Dutton thought the principle as important to geology as Newton's law of gravitation was to astronomy.

Thus, various geomorphological principles or laws were enunciated by the geological pioneers of the American West. But it was Davis who sought to link these ideas in a cyclic model for the Earth's geomorphic history. Apparently, he first formulated his ideas while working with Pumpelly in Montana for the Northern Pacific Railroad Survey when he noted that the Montana plains appeared to have been produced not from below sea level by elevation but by the wearing down of preexisting land of higher elevation, though there were also indications of rather recent uplift, as the watercourses seemed to be cutting down new courses (retrospective document by Davis written in 1933; in Chorley, Beckinsale, and Dunn 1973, 160). It

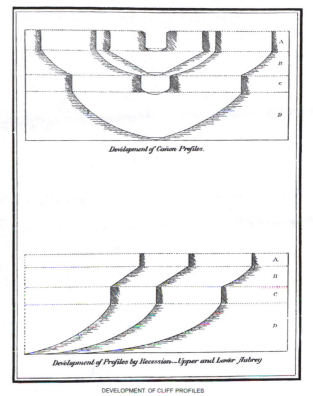

Development of Cañon Profiles.

Development of Profiles by Recession.—Upper and Lower Aubrey

DEVELOPMENT OF CLIFF PROFILES

Figure 7.1: Development of canyon profiles according to the hardness of strata (Dutton 1882, plate 40, facing p. 250)

was from this starting point and aided by reading the work of earlier investigators in the American west that Davis worked out his cyclic theory. His thoughts were profoundly influenced by evolutionary concepts and his writings permeated by organic metaphors. Indeed, his language was anthropomorphic to excess. Despite (or because of?) this, Davis's ideas dominated geomorphology in the late nineteenth century and during the first half of the twentieth.

Davis's grand idea was that landforms evolve or have histories analogous to those of living organisms: they have phases of "youth," "maturity," and "old age." His ideas were based in part on the developing notions of "laws" governing landscape development enunciated by the early geological explorers mentioned previously. But his cyclic ideas were first fully developed from his examination of the landforms of the Appalachians, nearer his home base at Harvard (Davis 1889b).

Davis invited his readers to conceive an area of land recently uplifted (ideally as a kind of plateau rather than as a steep-sided mountain range). Streams and rivers would be established on the new surface in a fairly random fashion if the subsurface structure were reasonably homogeneous. In this "youthful" phase, considerable portions of the original surface would still be present. The streams would cut into the land surface, widening their courses, so that divides formed between adjacent streams or rivers, until the original landform was largely obliterated. This was Davis's stage of "maturity." Finally, in "old age," the land surface becomes reduced once again to an almost level surface (or peneplain), rather like that of the original surface. So there was a *cycle* of erosion. Additional stages of "adolescence" and "senescence" could be added, and this was done on occasions. For a depiction of an idealized Davisian cycle of erosion, see Figure 7.2.

Moreover, if there were further uplift, either during a cycle or after its completion, then there could be a "rejuvenation" of the landform to something like the original "youthful" surface on which erosion could again get to work. Thus, there could be a series of complete or partial cycles, and it was the geomorphologist's task to analyze a landscape in such terms. One could display the history of a sequence of landforms (e.g., by looking for apparent planar, albeit incomplete, surfaces on contour maps), just as the geologist could reconstruct the tectonic, stratigraphic, and environmental history of a region (marine, lacustrine, desert conditions, and so on). Moreover, Davis supposed, by thinking in terms of landscape cycles, the geomorphologist's eye could be directed toward relevant phenomena. All this stood in the tradition of Hutton and Lyell in terms of ample geological time being available and processes that occurred in law-like fashion according to uniformitarian principles.

This doctrine fitted admirably with the spirit of Davis's times in that it was manifestly Darwinian. For example, in a preliminary statement of his ideas (Davis 1883, 357), it was suggested that, in considering the phenomena of antecedent drainage in the Appalachians, "the many pre-existent streams in each river-basin [have] concentrated their waters in a single channel of overflow, and . . . this one channel survives,—a fine example of natural selection" (or survival of the fittest!).

Davis's writing was wonderfully anthropomorphic. Rivers "lived and survived." Landscapes had "life histories" and "life cycles." A river was said to be "embarrassed" when it met structural inequalities that had to be "overcome." Rivers could "capture" other rivers like "pirates" or even "behead" them! The high plateaus of the Colorado River region were said to be "precocious." The process of

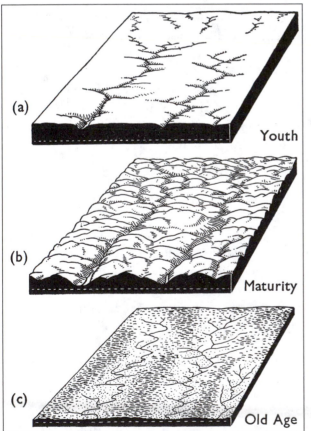

Figure 7.2: Evolution of a landscape according to Davis, from Holmes (1944, 187). Reproduced by permission of Taylor & Francis Ltd.

(a) Youth
(b) Maturity
(c) Old Age

erosion was analogous to a sculptor chiseling a block of marble. I particularly like the following passage (Davis 1889a, 108; emphasis added):

> The country hereabout [in Pennsylvania] was in ancient times a surface of faint relief, at a lower stand than now, traversed by *idle* streams; but in consequence of elevation to a greater altitude, the streams have *revived their lost activities*, and *set to work* to sink their channels and open out their valleys in the process of reducing the land to its *proper surface*, even with the sea; *for land finds its level*, like water, but more time is required before the level *is assumed*. The streams that drained the country when it was elevated *adopted such faint inequalities as they found for their first settlement*, and *have been engaged in perfecting their courses as best they could, cleaning them out*, deepening them, and *adapting them* most exactly to the *best* transportation of land-waste. In the processes of adjustment *thus called forth*, [with] every stream *struggling for its own existence*, it has sometimes happened that a stream with steep *head waters has seized the drainage area from the flat-lying head waters of an adjacent basin;* because, other things being equal, the waste of a surface is fastest on the steepest slopes [cf. Gilbert 1877], and hence the steeper streams have *gnawed more quickly* into the land-mass than the flatter ones, and the divide between a *pair of contesting streams* has consequently *been pushed* in the direction of the *fainter* descent.

From which could follow piracy and beheading!

All this was delightful albeit fanciful. Davis thought that such ideas would give geomorphology scientific status. And in a sense he was right, for his notion of cycles of erosion gave geomorphology its first general theory. The landform evolution of the Appalachians provided a model for subsequent thinking, interpretation, and research. Before Davis, the study of landforms might be said to have been in a preparadigm state (cf. Kuhn 1962) in that it had no unifying theory or modus operandi. The idea of peneplanation was especially well received, even if not everyone used Davis's anthropomorphic language. It should also be noted that Davis was a master at displaying his ideas about geomorphic histories by sequences of three-dimensional block diagrams that depicted changing structures and landscape forms. His artistic facility and charming literary style attracted generations of adherents.

Rather similar ideas were developed in Germany, notably by Albrecht Penck (1858–1945) in his *Morphologie der Erdoberfläche* (1894). (We shall meet Penck again in Chapter 9.) More generally, Davis's notion of cycling in the landscape was widely adopted by both geologists and geographers until well after World War II and persisted in textbooks even longer.

The Davisian scheme, developed in the wooded and well-watered Appalachians, needed modification when applied to arid or glaciated landscapes, and Davis subsequently extended his analyses to such landscapes and also to limestone topographies. However, his theory was challenged by Walther Penck (1888–1923), the geologist son of Albrecht Penck. As one who did considerable work in tectonically active areas such as Argentina and Turkey, Penck Jr. doubted that things could ever be as simple as Davis's model suggested.

It was inadmissible to suppose that uplift occurred suddenly at first and was then followed by erosion separately. The two processes would occur together. And the Davisian scheme could hardly work if a land surface were constantly "bobbing up and down."

Working particularly in Germany's Black Forest region, Penck Jr. tried to develop a model that would take account of earth movements and erosional processes proceeding simultaneously. Moreover, there could be broad foldings due to lateral pressures, with the formation of synclines accompanying the formation of anticlines—giving the beginnings of "basin and range" topographies, or regional arching. In addition, the two kinds of earth movement could occur together, as in the Alps. Since the rate of uplift could vary (perhaps starting slowly, accelerating, and then decelerating) the down-cutting by rivers might or might not keep up with the earth movements. There could also be relevant climatic changes. Things could be much more complicated than Davis supposed. In Penck's terminology, he sought to provide a "morphological analysis" (Penck 1924) rather than some standard anthropomorphic model.

Writing on his Black Forest studies, Penck (1925) envisaged the erosion of a rising and expanding regional dome, with upward movement accelerating. The observed series of benches was supposedly produced under such circumstances, but benches did not mark periods of low erosion due to pauses in the uplift. Penck was moving away from the cycle of erosion (youthful, mature, and aged landforms) to the idea of waxing and waning erosion, with the former producing convex-shaped inclined flats and the latter yielding concave-shaped inclined surfaces. Benches, he thought, could be produced without pauses in uplift.

Davis (1932) responded, somewhat misrepresenting Penck's ideas in the process (see Chorley, Beckinsale, and Dunn 1973, chap. 28), suggesting that Penck's convex and concave surfaces referred to valley sides rather than inclined flats. But Davis's thinking changed nonetheless to the view that a land surface doesn't necessarily flatten out as it recedes by erosion, with the slope gradually decreasing. It may be worn back, maintaining a fairly level erosion surface, together with one of greater declivity at the locality where the main erosive action is occurring. That is, slopes can recede laterally, maintaining an approximately constant declivity. And, Davis acknowledged, this would be the way things occur in arid climates, with the formation of "pediments" (gently sloping rock surfaces in front of a relatively abrupt receding scarps) rather than peneplains. Such slow erosive processes in desert country could lead to the production of isolated hills "sitting on" "pediplains." Ayers Rock in Australia is a famous example.

Unfortunately, but unsurprisingly, communication between the English- and German-speaking world was not good in the 1920s, and American (led by Davis) and German geomorphologists (led by the Pencks) began to drift apart, the separation being accelerated by World War II. So, according to Chorley, Beckinsale, and Dunn (1973, 507): "By World War II the Davis–Penck controversy, as it was carried out in the English-speaking world, had foundered in a doctrinaire and depressingly-semantic morass"—which we shall be prudent to avoid sinking into here!

It is probably fair to say, however, that it was the Germans who "won" in the long run. Russell (1958, 2) has put the story well:

> The concept of the erosion cycle accounted for a rash of peneplain hunters who were likely to regard anything from alleged accordance of summit levels to broad alluvial flats as evidence of a completed [erosion] cycle. The quest eventually lost popularity, however, so that within recent years the rate of peneplain discovery has come to a near halt. . . .
>
> Peneplain enthusiasts commonly disregarded the effects of isostatic compensation. They failed to recognize the proposition that reduction in actual relief is many times slower than the rate of denudation of a rising rock column, so that sharply crested ridges and deep valleys [can] persist for enormous intervals of time.

Rejection of Davisian cycles, pure and simple, has continued. In 1998, Paul Larson posted a query on the Internet (http://main.amu.edu.pl/~sgp/gw/wmd/wmd.html), asking how many people accepted Davis's cyclic model of landscape evolution. The response was overwhelmingly negative. One respondent wrote, "There's no place I know in the world today where a long Davisian cycle can be demonstrated." Another said that geomorphologists have to consider climates and processes changing through time, the lithological control of landforms, and so on. Another thought that the theory was of limited value in tectonically active regions and showed both Darwinian and Marxist(!) influences (because it proceeded in distinct stages?). Indeed, the theory was a "complete turkey" when applied to landscapes such as those in Australia. Another respondent thought (on the basis of examining a range of textbooks) that one could say that the Davisian cycle of erosion concept itself passed through four stages: (1) presented as a fundamental concept, (2) described but also criticized, (3) mentioned for its historical interest as a step in unifying geomorphology, and (4) no mention made of either Davis or geomorphological cycles. And so theories come and go. One might even say that the wheels fell off Davis's cycle!

8

MOUNTAINS, BASINS, AND SEDIMENTS

There was something insufficient about the "Huttonian" theory of Lyell as to the random rises and falls of the surface of the Earth at different times and places. What caused such changes? They were still not explained.

In Chapter 6, I mentioned Cuvier's catastrophist theory. It was his previously mentioned countryman Léonce Élie de Beaumont (1798–1874), professor at the *École des Mines* in Paris and Cuvier's successor in the chair of natural history at the *Collège de France,* who put forward a theory to account for the sudden worldwide episodes of destruction of life that Cuvier thought were indicated in the stratigraphic record.

In his early career, Élie de Beaumont worked on the preparation of a geological map of France. In 1829, while engaged in this work, he delivered a paper setting forth an initial exposition of his ideas on tectonic activity and sought to relate this to the stratigraphic record that he was studying for his map work (Élie de Beaumont 1829–1830). The theory proposed that the Earth had a hot interior, but it was not permanently hot: it had been gradually been losing heat since its formation. The crust had by now reached an approximately constant temperature, but cooling of the molten interior continued and was thus accompanied by *contraction*. So from time to time, the solid crust buckled and formed mountain ranges. The buckling was not thought to have occurred uniformly all over the globe at the same time, but where mountain ranges were parallel it was assumed that the tectonic agencies producing them occurred at about the same time. That is, the parallel ranges were associated *"formations."*

But the uplift of mountains in a tectonic episode (today called an orogeny) would also be accompanied by the formation of marine basins and worldwide changes in sea level. Sediments would also be deposited (or "formed") in the basins, so marine sediments might be *correlatable* worldwide. Thus, the theory had both a plausible physical cause and an explanation of the stratigraphic evidence. Moreover, after the formation of a mountain range (or a contemporaneous set of associated mountain ranges with similar alignments), additional cooling could occur, and the process might be repeated, though the next set of mountains would have a different alignment from that of the first. Then it could happen again. And again!

All this might make it appear that Élie de Beaumont's theory was of the captain and mate variety. There were, indeed, repeated tectonic cycles. However, the Earth's configuration changed after every episode of mountain building, to the point that Élie de Beaumont (1852) fancifully supposed that mountain ranges eventually formed a regular network of pentagons—or a *réseau pentagonal,* as he called it (Figure 8.1).

For Élie de Beaumont (1852), there were associated patterns for the Earth both in space and in time. He imagined that, ideally, a zone of mountain formation could form a half circle around the Earth. To the present, 15 such semicircles had been formed, intersecting with one another so as to form a network of 12 pentagons, making the form of the world's mountain ranges dodecahedral. The theory was both cyclic and linear. On the one hand, there were repeated episodes of contraction and crustal crumpling. On the other, the history of the Earth was proceeding in one direction: to a smaller, cooler, more "wrinkled" globe.

Over the years, Élie de Beaumont's theory became increasingly "baroque." It started in 1829 with only four "systems" or phases of mountain building (and associated basin formation and sedimentation). By 1833, there were 12, and by 1852 there were 15. He even wrote to a colleague (Constant Prévost [1787–1856]) that he thought that there *might* be more than a hundred! The theory was becoming unfalsifiable, for additional circles of mountains could be added endlessly to patch up the theory (by ad hoc hypotheses). It should be remarked that the "wrinkled-apple" theory was by no means implausible, but the suggestion that the successive wrinkles should eventually yield a regular geometrical pattern outran the evidence. Nevertheless, Élie de Beaumont's theory, aided by his powerful position in French science, had adherents in France and in Britain in the 1830s (see the discussion later in this chapter), but thereafter it was largely a French theory. Of course, as the supposed number of "formation" episodes increased, the theory became more "gradualist" than "catastrophist."

Élie de Beaumont's theory had other interesting aspects. If we look at Hutton's section of Arran (Figure 5.2), we see three main sets of rocks: a central granite mountain, adjacent strata apparently bent up by intruded granitic magma, and strata lying unconformably over them, mostly horizontal but somewhat upturned near the contact. Considered through the lens of Élie de Beaumont's theory, we have two sets of basin sediments situated on either side of the mountain, some of which have been curved upward by crustal crumpling and intruding magma. The almost horizontal sediments at the furthest right and left of the picture (especially at the right) would be the youngest. In regard to such a case (though *not* specifically this one), Élie de Beaumont wrote (1831, 242; emphasis added),

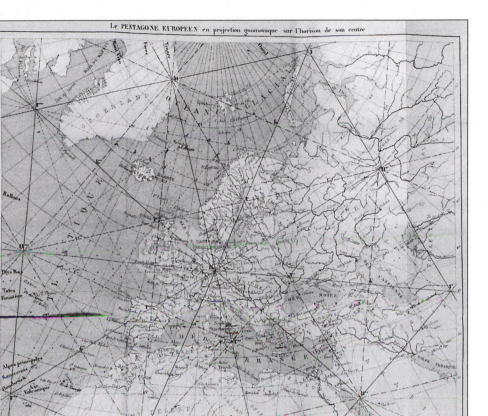

Figure 8.1: Part of the Earth's pentagonal arrangement of mountain ranges (Élie de Beaumont 1852, vol. 3, plate 5). Reproduced by permission of the British Library.

At the same time that the position of the ancient and inclined beds furnishes the best proof of the elevation of the mountains of which they constitute a part, the geological age of these beds affords the best means of determining the relative age of the mountains themselves; for it is evident, the first appearance of the chain itself is necessarily *intermediate between the period* when the beds, now upraised, were deposited, and that when the strata were produced horizontally at its feet.

Thus examination of *structures* could give an idea of relative *ages*.

However, for Arran, the granite mountain was essentially isolated, not a chain, and the peripheral basin sediments are not represented as exactly horizontal in Clerk's drawing. Moreover, Élie de Beaumont's mountain chains were not necessarily granitic. They could also be ranges of strata, folded more or less gently. He gave an example from Adam Sedgwick's (1785−1873) work in northern England. The Lake District mountains ("Transition rocks" in Wernerian terminology; Ordovician−Silurian in modern terms) ran west-southwest to east-northeast, while the Pennine range to their east (Carboniferous) ran north to south. Strong faulting occurs where the two ranges come in contact. Élie de Beaumont (1831, 245) thought the movements that inclined the older Transition rocks must have occurred between their deposition and that of the nearly horizontal Carboniferous rocks (i.e., during Old Red Sandstone time). He imagined that the movements of the Transition rocks were "catastrophic," for Sedgwick had found them to be fractured by faulting. Similar arguments were suggested for European rocks, such as in the Vosges of Belgium.

Another Élie de Beaumont idea was that early in Earth history, the rocks of the crust were more heated and altered by the adjacent material of the planet's hot interior and thus were highly altered or metamorphosed. Later, the heat had retreated, so to speak, into the Earth's interior, and thus the alteration would have been less intense, except around intruded magma masses. So he thought that the *type* of metamorphism would indicate its *age*. This—and the idea that mountain ranges of similar orientation were of similar age—was false, but the ideas were influential.

The stratigrapher Constant Prévost, a founder of the *Société Géologique de France,* was a contemporary of Élie de Beaumont and a former student of Cuvier and his coworker Alexandre Brongniart (1770−1847). Prévost (1839−1840) took issue with Élie de Beaumont's theory, leaning toward uniformitarian geology. Prévost had worked in the Paris Basin, where Brongniart and Cuvier had suggested an alternation of terrestrial and fresh- and saltwater conditions (with corresponding changes in fauna). But Prévost pointed out that fossils might have been eroded out and redeposited, so that things might not be as they appeared. In fact, Prévost thought the evidence suggested a progressive regression of the sea through the Earth's history (as compared with Cuvier's successive inundations and regressions). Volcanoes were not uplifted "blisters" on the Earth's surface (the so-called craters of elevation theory) but accumulations of extruded ash or lava, as Lyell envisaged.

But, with Élie de Beaumont, Prévost liked the idea of a cooling and contracting Earth. Basins were formed where contraction was greatest, leaving the mountains standing higher. On this view, no special mechanism was required to produce uplift. This was hardly a model that would, in itself, lead to a continuing marine regression, and it was not, of course, cyclic. But it did support the tectonic doctrine that dominated well into the twentieth century—a cooling and contracting Earth. This culminated in the great synthesis of Eduard Suess, to be considered later in this chapter. The scene now shifts to North America, where, as we have seen, great advances were made through the western surveys and important ideas on geomorphology developed. Brief mention has also been made of Dutton's idea of isostasy (p. 70).

The leading American geologist in the mid-nineteenth century was James Dwight Dana (1813−1895), professor at Yale from 1850 to 1890. In his early career, he was a naturalist aboard the naval expedition (1838−1842) commanded by Charles Wilkes. This voyage of some 87,000 miles took the investigators over much of the Pacific, both to its islands and along the western side of the United States. A suitable position for the western end of the border with Canada was proposed, and some inland exploration was undertaken in the northwestern United States. Additionally, the areas of the southern oceans were charted, and it was established that Antarctica was a continent. Australia and New Zealand were visited, with Dana making important observations in both countries. From such experiences, Dana, like Darwin, acquired much scientific information that he used for subsequent theorizing. But whereas Darwin's thoughts moved in the direction of evolutionary biology, Dana's trajectory led toward mineralogy and a synthetic theory of the Earth. Dana was a religious man and a "progressionist" geotheorist. He even saw the hand of God supervising the geological expansion of North America!

During the Wilkes expedition, Dana became interested in coral reefs. Some Pacific islands, like Tahiti and Hawaii, had volcanic centers, and a few had active volcanoes. Some islands formed chains, with active or more recent volcanoes at one end and indications of the islands' subsidence at the other end; Dana even predicted the existence of drowned atolls (see p. 101.) There were also the arcuate volcanic island chains bounding the Pacific, apparently occurring in areas of uplift.

On his return to the United States, Dana began a notable series of papers (mostly in the *American Journal of Science,* which he edited from 1846) and books (especially *Geology* [1849] and *Manual of Geology* [1862]) in which he formulated a synthetic view of Earth history and the way the planet "worked" or had evolved since its formation. Unlike the Lyellians, he began to include the idea of *lateral* earth movements. Evidence for lateral movements was coming from observations in Switzerland in the 1840s, but Dana was more influenced by the observations in the Appalachians of the brother geologists William and Henry Rogers (1805−1881, 1809−1866), where the form of the folds suggested lateral forces.

Like others, Dana (1847a, 1847b) envisaged the Earth as cooling and contracting. Irregular contraction in early times had, he supposed, produced large depressions that now formed the ocean basins. The regions left standing high were thought to be the "nuclei" of the present continents. So, in the last analysis, mountains were produced not by uplift but by withdrawal of ocean waters into deepening troughs. Thus, the continents and oceans were essentially permanent features of the Earth's form. In this manner, Dana dealt with the old Hutton/Lyell problem of uplift. Dana's idea of the permanency of the ocean basins lasted in one form or another until the second half of the twentieth century.

Like Élie de Beaumont, Dana (1847b) supposed that most mountain ranges were elongate. But they could be curved (though not as great circles) or straight, and they might intersect. They were supposedly associated with former fissures in the Earth's crust and represented lines of equal cooling and contraction, where associated tensional forces had been generated. Dana rejected Élie de Beaumont's idea that the ages of mountain ranges were related to their alignments. But according to Şengör (2003, 111) Dana was nevertheless influenced by the Frenchman. Both thought of global tectonics by reference the appearance of the Moon, and both were contractionists. For Dana (1846a, 1846b), parts of the Moon had craters (supposedly volcanic), while other areas appeared relatively flat. This suggested that volcanic activity (and cooling) had begun earlier in the smoother areas than in the regions with numerous craters. Similarly, the central plains of North America might be cooled-off areas, whereas areas like the Pacific were still volcanically active and hence cooling and contracting.

Then Dana (1847a) sketched a series of pictures showing his ideas about how the ocean basins might have been formed by global cooling. The result of the ongoing sequence of events (Figure 8.2) was that the oceans progressively deepened over time, leading to marine regression worldwide, which meant that there was *enlargement* of the continental areas through geological time. Moreover, the tension set up between the contracting oceanic crust and the previously cooled and stabilized continental crust could lead to occasional catastrophic ruptures in the form of major faulting near the ocean−continent margins, accompanied by folding of strata to form mountain ranges at these margins. Volcanic activity could occur along the seaward sides of the ranges.

Dana added to his theory in his Presidential Address to the American Association for the Advancement of Science (1855) and in a paper published the following year (Dana 1856) with the idea that continents grew, or successively enlarged, from nuclei or "germs." In the North American case, the "Azoic nucleus" was

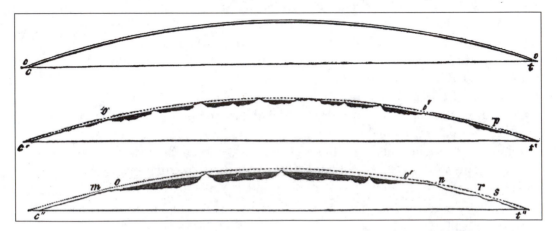

Figure 8.2: Cooling, contracting globe (Dana 1847a, 181). In Figure 1, the Earth's primeval envelope of water is shown along the curve o−o′. Figure 2 shows the globe after cooling and contracting, with incipient ocean basins formed. In Figure 3, it is supposed that, as it contracts and subsides, the contracting solidified crust in the central area exerts lateral tensional forces on the adjacent, previously solidified crust.

located in what is today called the Canadian Shield, or the area around Hudson Bay. Successions of accretionary growth occurred in V-shaped form to the east, south, and west of this "germ," in time forming modern North America (Dott 1997). Dana related the supposed directions of lateral forces acting on the western and eastern sides of the continent to the foldings and cleavages of the rocks observed of the western (Rocky Mountains) and eastern (Appalachian and so on) ranges. In the successive accretions of land to the continent, we can see a kind of cyclic action at work (for more on this, see p. 103–104).

Another major contributor to American theorizing in the mid-nineteenth century was the paleontologist and stratigrapher James Hall (1811–1898) of the New York Geological Survey and later the Iowa Survey and first president of the Geological Society of America. (He was also associated with several other state surveys.) Hall was not a "cyclist" per se, but it is appropriate to refer to his ideas here, as he exerted such an influence on the history of ideas about mountain building. In an address to the American Association for the Advancement of Science in 1857 (only formally published in 1883; but see Hall 1859, 1860; Hall and Whitney 1858), Hall expressed important ideas about the origin of mountains. He did not accept the theories of Élie de Beaumont and Dana. Rather, he was influenced by the English astronomer John Herschel.

In a forerunner of isostasy theory (see p. 70), Herschel (1837) thought the Earth's crust and subcrust were plastic, slowly responding to pressures and tensions. So if an area accumulated sediment, the crust might be depressed, and more sediment could be deposited there. Conversely, erosion of sediment could result in upward movement. Hall adopted this idea and applied it to the Appalachian region, where there were enormous thicknesses of inclined sediments—much thicker than strata further west toward the continent's interior—in an area later called a "craton" (see p. 159) or a stable part of the Earth's crust. These Appalachian sediments, thought Hall, could not have been deposited into a preexisting trough of fixed size: the sea could not have been deep enough to accommodate the sediments. The region to the west of the trough, with fossil corals, was evidently shallower.

So Hall had the idea that a down-warping region of the Earth's crust, loaded by sediment deposition, would lead to the sediments in the "trough" being folded by compression in their upper layers and fractured in their lower layers, where the cracks would be intruded by magma from below. For the ranges of New England, sediment was supposedly transported into the trough from northeast to southwest, in the same direction as its alignment, and approximately parallel to the margin of the interior (or craton). Insofar as there was elevation, it was due to plication of the upper layers of the sediments in the trough. There was evidence (such as mud cracks) for some of the sediments having been deposited in shallow waters. Thus, being 40,000 to 50,000 feet thick, there *had* to have been subsidence.

But Hall had little to say about how the sediments might have been converted into mountains. He was interested in the deposition of the enormous thickness of sediments and the fossils they contained and thought there was causal association between the huge thickness of sediments and the mountains in which they now occurred. But he wasn't really trying to develop a theory of mountain *formation,* though that is how he was interpreted, for the world viewed him through the lens of Dana's response to Hall's work (for more, see Şengör 2003, 123–30).

Dana couldn't accept that Hall offered a satisfactory theory of mountain formation (and neither can we). Dana thought that a subsiding sediment trough would require a thin, plastic crust with liquid magma below. And were this so, it would be difficult to see how it could hold up large mountain ranges. As he famously remarked (Dana 1866, 210),

> Mr. Hall's hypothesis has its cause for subsidence, but none for the lifting of the thickened sunken crust into mountains. It is a theory for the origin of mountains with the origin of mountains left out.

Nevertheless, Dana was sufficiently stimulated by Hall to return to the question of the origin of mountains in a later series of papers in the 1870s (Dana 1873a, 1873b, 1873c, 1873d, 1873e). In Dana (1873b, 430), we get a new term: "geosynclinal," meaning a trough-like area of down-warping, which can receive or has received large quantities of sediment over long periods of time—an idea obviously beholden to Hall. Its converse, a "geanticlinal," was a large upwardly bent area of crust. Such structures could supposedly form in response to lateral compression, producing, over time, a succession of basins and their sediments and growth of continents at their margins in a kind of cyclic fashion. A reconstruction of Dana's theory of 1873 has been given by Dott (Figure 8.3). There was a kind of cyclicity to the theory in that there were repeated episodes of continental growth (an idea carried over from Dana's earlier thinking). Moreover, continents "expanded"—in a manner congenial to nineteenth-century "nation builders"!

Subsequently, the word "geosyncline" was substituted for "geosynclinal." The lateral forces of Figure 8.3 supposedly arose from the Earth's contraction as it cooled. The "wrinkled-apple" theory was still at large. But the uplift issue remained obscure, and the theory didn't seem to account for all mountain ranges, especially in the European Alps.

In the event, it was from Europe that a master geologist emerged to offer an all-encompassing theory of the Earth: Eduard Suess (1831–1914), professor of geology at Vienna University. Suess's principal early work was concerned with the paleontology of graptolites, trilobites, brachiopods, and mammals. Subsequently, he worked on the complex geology of the Alps and correlated the Triassic, Jurassic, and Cretaceous strata there with equivalents outside the Alps. He used Cuvier's "comparative method" in paleontology (i.e., utilizing analogies between modern and fossil forms to reconstruct the skeletons of extinct animals) and began to do the same in stratigraphy. Thus, he studied Italian earthquakes and their general distribution, concluding that there was an association between earthquake activity and mountain formation. So, in 1875, he published his first book, *Die Entstehung der Alpen (The Origin of the Alps).*

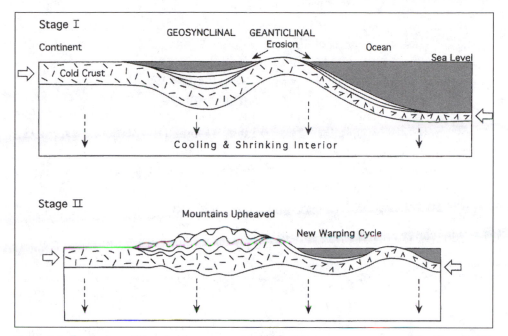

Figure 8.3: Depiction of Dana's ideas on mountain building, as reconstructed by Dott (1997, 299). Reprinted by permission of the *American Journal of Science*.

In this volume, obviously influenced by what could be *seen* in the Alps, Suess argued for the importance of *lateral* forces in their formation, apparently acting from south to north. It seemed as if the strata had been pushed toward and partly over a stable and resistant "foreland" *(Vorland)* or platform (later the "craton" of Stille), with a convex northern margin. On the southern side, there was a roughly concave margin and a "backland" (*Hinterland,* or "behindland") of lower altitude where volcanic and seismic activity occurred (in Italy and the Mediterranean area). It seemed that there was compression to the north and tension to the south, where fractures in the Earth's crust caused the volcanism. This meant that, insofar as there was uplift, it was not due to the upward pressure of magma per se. Rather, molten magma penetrated the crust via tension-induced fractures.

In the last chapter of *The Origin of the Alps,* Suess sought to subsume these proposals under the general suggestion that the forces were produced by the Earth's cooling and contracting, from time to time producing great episodes (cycles!) of mountain building (later called orogenies). He also began to think globally in that he recognized the "biosphere" *(Biosphäre),* to be set alongside the lithosphere, hydrosphere, and atmosphere when thinking about the Earth. Additionally, he introduced the famous phrase *Das Antlitz der Erde* ("the face of the Earth") as the title of his three-volume magnum opus (Suess 1885−1909) (also French [1897−1918] and English [1909−1924] translations).

The phenomenon of land elevation in Scandinavia (recognizable by the occurrence of shorelines well above sea level) also attracted Suess's attention. Initially, he supposed that it could be accounted for by his theory of contraction (with a north-directed movement induced by contraction, not upthrust from below) (Suess 1875, 151). But in *The Face of the Earth,* he developed the idea that the elevated strandlines could be attributed to falls of sea level or that they were the marks of former ice-dammed lakes. He never accepted that there had been independent uplift in Scandinavia. (Today the uplift there is ascribed to the melting of the ice load after the Ice Ages.)

The general idea of Suess's theory, as presented in *The Face of the Earth*, worked in accordance with one of his most famous statements: "[t]he breaking up of the terrestrial globe, this it is we witness" (English translation, 1904–1924, 1:604) or "what we are witnessing is the collapse of the terrestrial globe." Like several predecessors, Suess believed that the major agency producing mountains and ocean basins was global cooling and contraction. Given the contraction of a fluid interior and the existence of a solid crust, there were episodic crustal ruptures. But the result was not necessarily a regular geometrical structure of intersecting mountain ranges such as Élie de Beaumont supposed. Suess emphasized that many mountain ranges had curved forms. There could be a great variety of folds and faults (especially overthrusts, such as in the Glarus Canton, Switzerland, recognized since the 1840s), or collapse structures might be formed, such as Gilbert and Powell had remarked in the American West (Figure 8.4[a]).

So the Earth's evolution was a one-way process, and every time an event such as that implicit in Figure 8.4(b) occurred under the sea, there would be a *global* lowering of sea level. The figured example would not produce a dramatic lowering of sea level, but Suess supposed that there could be much larger-scale collapse events, giving rise to the major ocean basins. These approximately elliptical basins might come to intersect one another, producing the outlines of the several continents tapering to the south. Moreover, lateral compression could elevate mountains, without force being exerted upward from within the Earth.

Then, following some large-scale subsidence—which also produced

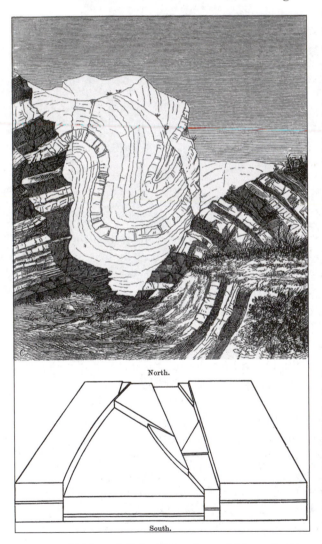

Figure 8.4: Folds and dislocations due to (a) lateral compression (b) subsidence, according to Suess.
(top) Suess (1904–1924, 1:108): folded shales in Moravia.
(bottom) Suess (1904–1924, 1:132): collapse structures in the Wasatch Plateau.

mountains as a result of the contraction—erosion and sedimentation would get to work, and the marine basins would be filled, giving rise to *worldwide* marine transgressions. So there was an *explanation* of the correlations that could be made for the various parts of the stratigraphic column in different parts of the world of approximately the same age—why, for example, the Permian can be recognized as such in quite separate parts of the world. Here was a powerful geological theory!

Suess coined a new term for the worldwide changes in sea level that he envisaged: "eustatic changes." As he put it (English translation, 1904–1924; [emphasis in the original]),

> *The crust of the Earth gives way and falls in; the sea follows it.* (2:537)

And

> *The formation of the sea basins produces spasmodic eustatic negative movements.* (2:538)

The phenomenon of global changes in sea level following from spasmodic tectonic events came to be called *eustasy,* though Suess himself did not employ this term as a noun. A general lowering of sea level he called a "negative" eustatic movement, while a rise was said to be "positive."

The collapses could be regarded as large-scale but spasmodic tectonic events and in Europe were called the Caledonian (in the Silurian), the Variscan (or Armorican or Hercynian) (Permian/Carboniferous), and the Alpine (Tertiary) orogenies. The oldest occurred in northern Europe (Scotland and Scandinavia), and the youngest occurred to the south in the Alps. Each involved mountain building, metamorphism, seismicity, and volcanism and was followed by sedimentation and the emergence of new life forms. Thus, there was a grand cyclicity in Suess's theory. And each of the main geological systems (Devonian or whatever) appeared to close with a rise in sea level. The stratigraphic column thus began to have a theoretical rationale, suggested as follows:

> *The limits of the formations established by William Smith and his successors [in England and Wales] correspond for the most parts with [the] negative phases [of eustatic movement].* (Suess, English translation, 1904–1924, 2:541)

The tectonic changes served as a kind of "motor," driving biological evolution (but Smith was not an evolutionist).

At the end of his book (Suess, English translation, 1904–1924, 4:chap. 18), Suess discussed the remains of four ancient landmasses that had supposedly escaped collapse (Dana-esque talk): Laurentia (named after the St. Lawrence River in Canada, in the region of the Canadian Shield), Angaraland (the tableland of East Siberia up to the Arctic and perhaps also parts of China), Gondwána (the Indian peninsula, Madagascar, Africa from the south of the Karoo to the Sahara, and much of Brazil and Patagonia), and Antarctis

(Antarctica with Australia and Patagonia). To this list we might add what we call the Baltic Shield, which Suess did in fact discuss in chapter 2 of his volume 2 (English edition). The former existence of these lands was based partly on paleontological evidence. For example, characteristic fossil plants (the *Glossopteris* flora) were found in the several parts of Gondwána.

Between India and Africa (now supposedly separated parts of Gondwána), it was suggested that younger mountains (e.g., those of Asia Minor and the Middle East) had emerged from an ancient Mediterranean ("mid-Earth") sea, which Suess called the "Tethys," after the ancient Greek sea goddess of that name. Young mountains were also found encircling the Pacific Ocean.

Suess called his four relatively undisturbed, ancient regions "places of refuge" or "asylums" untouched since Carboniferous times—"tracts in which terrestrial forms of life have been protected from the action of such physical changes as [marine] transgressions and mountain building, for a very long period" (English translation, 1904–1924, 4:660).

Suess's discussion of stable protected regions was by no means wholly satisfactory. He knew of Gondwána fossils in Australia, yet he placed that continent with Antarctica rather than with, say, India. (But there is, in the modern world, an "ancient" flora of tree ferns and so on found in Australia and New Zealand of typical southern hemisphere type.) And Gondwána itself had seemingly been broken into by the Atlantic Ocean. So the stable asylum was not so very stable! Suess (English translation, 1904–1924, 4:666) realized that there were similarities between West Africa and eastern South America. So he supposed there might have been a "possible continental junction." This did *not* imply continental drift for Suess but rather the collapse and submersion of part of Gondwána.

Suess's ideas were based on an encyclopedic knowledge of the geological literature, worldwide, in a variety of languages. His erudition and influence were tremendous. But, while he rejected Dana's notion of geosynclines and created a *general* theory of the Earth and its history, based on the plausible contraction theory, when examined closely his model of mountain building was dubious, as we see from its diagrammatic representation by Şengör (Figure 8.5).

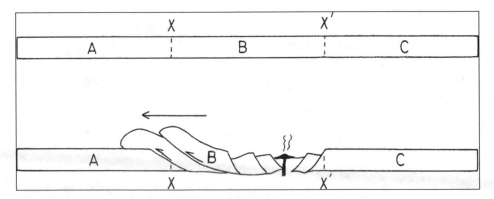

Figure 8.5: Suess's model of mountain building (Şengör 1982, 400, fig. 4). Reproduced by permission of Springer-Verlag and Professor Şengör.

Suess did refer to isostasy ("equal standing") in the last volume of his magnum opus, but his theory was developed before Dutton (1892) enunciated the concept, so Suess did not use it much in his theorizing. But isostasy is most important for our theme in that it adds an extra "dimension" to understanding the rises and falls of land and sea levels and hence of geological cycles of various types. Isostasy suggests that as ocean basins are filled with sediment, their floors are depressed by the weight of the added material. And as erosion reduces mountain ranges, they "lose weight" and may consequently be elevated to maintain isostatic equilibrium. Such up-and-down adjustments would have appealed to Lyell, though they were not the result of magma pushing up the crust from below, as he envisaged (though we think that the comparatively light rock granite can rise by buoyancy).

Suess's ideas were developed in the United States by Thomas Chamberlin (1843–1928). He initially worked with the U.S. Geological Survey, was later president of the University of Wisconsin, and completed his career as chairman of the Department of Geology at the University of Chicago. His interests ranged widely, but his specialty, especially while with the Survey, was glaciology, and he had an interest in the factors influencing climate or changes thereof.

By the end of the nineteenth century, doubts were being expressed as to whether global contraction could account for the lateral deformations evident in mountain ranges. Hence, Chamberlin sought to replace thermal contraction by the agencies of gravity and isostasy as "motors" for geological change. He also utilized the concept of base leveling developed by American geomorphologists (Chapter 7). Like Suess, he sought a theory to account for the major subdivisions of the stratigraphic column, with worldwide lithological and paleontological changes accompanying worldwide unconformities separating the different geological systems (Silurian, Devonian, and so on). It was a grand project!

Chamberlin also wrote about the Earth's origin, suggesting that it had formed from the accretion of a cloud of "planetesimals," or small, cold bodies, which, as they collided and adhered, give rise to the Earth and generated its heat as a result of the impacts. But he did not utilize ideas about different internal regions of the Earth (some molten, some not) that were being developed by his contemporaries as a result of seismological investigations (which we leave aside here).

As Chamberlin's hot Earth cooled and contracted, some parts supposedly did more so than others, producing thereby ocean basins and continents. Chamberlin spoke of the regions of the former "wedging" down into the Earth. But because the area of the oceans was so much greater than that of the continents, the latter were elevated and squeezed in the process, producing the characteristic faults and folds of mountain regions, eventually reaching a condition of isostatic adjustment. There followed erosion and deposition of sediments in the basins, raising the sea level, and the increased pressure in the basins tended to elevate the continental areas.

Chamberlin further supposed that there were periodic founderings of the ocean basins, with, of course, concomitant elevations of the continental areas. Following an episode of land "elevation," erosion and base leveling would occur, with the seaward extension of the "continental shelves"—the regions of fairly shallow water near land, with a sudden declivity to deeper water further offshore. (The existence of such submarine topography was known from sea-floor soundings in many parts of the world by the British navy's research vessel *Challenger* in the years 1872–1876.) If there were a further basin subsidence, the sea level would again fall, taking the shoreline close to the edge of the continental shelf. And thus the area of shallow water—an important marine habitat—would be reduced, and the ecological change would be a stimulus to evolutionary change. Such changes would be essentially global in scope.

Chamberlin further suggested that when rocks were weathered, the chemical changes would consume atmospheric carbon dioxide and that there could be global cooling because of a reduced "greenhouse effect." Conversely, during a marine transgression such as occurred during the Cretaceous, there would be more carbon dioxide in the atmosphere and a warmer climate worldwide (with melting of the ice caps). So there could be changes in climate accompanying the tectonic changes as well as changes in the areas of shallow ocean waters. If all this occurred episodically, then there was a "basis for the strict correlation of transoceanic action and for the *division of geological history into its natural epochs* [or systems]" (Chamberlin 1898, 461; emphasis added). Chamberlin (1909) went further, suggesting that base leveling, as he envisaged it, implied the formation at any given time of a homologous series of deposits worldwide. So there could be worldwide correlations of unconformities (even if different parts of an unconformity might not be of exactly the same age since erosion and marine transgression both required time).

The major (tectonic) geological changes, postulated to occur worldwide because of ocean-basin founderings, Chamberlin called "diastrophism," which he supposed (1909) was the "ultimate basis of [geological] correlation." This was an important desideratum at the time, given that the early meetings of the International Geological Congress (first held in 1878) had been called to sort out the mess developing as different countries established their separate and often incompatible stratigraphic columns, and they had not achieved notable success. If someone (like Chamberlin) could provide a rationale for a global stratigraphy, it would be most welcome.

Not all geologists accepted Chamberlin's ideas, however. At any given time, different deposits could be formed in different parts of the world: clays here, sands there, limestones somewhere else. Further, Johannes Walther (1860–1937) in Germany wrote, "It is a principle of far-reaching importance that only the facies [i.e., the general appearance and characteristics of a rock unit corresponding to some particular mode of formation] or facies areas that are at present adjacent to one another can be geologically superimposed upon one another" (Walther 1894, 2:979). In other words, the distribution of sediments deposited *horizontally* at the *same time* will have an analogous *vertical* distribution at some locality, having a *chronological order*. For example, if sands are deposited near a coast and finer sediment is carried further offshore and deposited simultaneously and if the marine shelf is extended outward by ongoing sedimentation, then a particular locality that has acquired mud will in time receive sand as the shoreline moves outward. Mud thus gives way to sand both laterally and vertically.

Thus *exact* worldwide correlation could not be expected. Nevertheless, Chamberlin set the scene for the great interest in sedimentary cycles that developed in the twentieth century (see Chapters 10 and 11).

We have now said something about the development of ideas about global contraction, proposed in the nineteenth century, and a little about the development of such ideas in the twentieth century. But, as all high school students "know," that was all wrong! The "correct" idea is that which was developed in the 1960s and eventually accepted for the most part in the 1970s, becoming a new "paradigm" for the Earth sciences. I refer to the doctrine of "plate tectonics," according to which the Earth's crust consists of a number of "plates"— some above sea level, some below, and some both on land and under the sea—which move around relative to one another, their motions being driven by convection currents within the Earth's interior.

To digress for a moment: through seismic work, the Earth is known to have a solid inner core, a fluid core, a large rocky but hot and somewhat plastic "mantle" made chiefly of the mineral peridotite, and a relatively rigid "crust." The mantle is sufficiently plastic to allow slow-moving convection currents that drive the movements of the overlying crustal "plates." These "ride on the back" of the moving substratum, so to speak, so that the plate motions are driven by the internal convection currents. The physical state of the internal parts of the Earth is determined by the "balance" between temperature and pressure. The upper part of the mantle—called the asthenosphere—is more plastic than the deeper material, which is at a higher pressure. This plasticity allows the "floating" of mountain ranges on a quasi-fluid substrate, making possible the isostatic adjustments, referred to previously. The existence of a fluid "core" is evident from the fact that it cannot transmit transverse seismic vibrations and thus produces a "shadow" for earthquake-generated transverse waves (Oldham 1906). The dimensions of the liquid core can be calculated from the size of the "shadow." There is more active convection in the core, whence arises the Earth's magnetic field (see Chapter 12). The term "mantle" derives from the German word *Mantel*. It was introduced by the seismologist Emil Wiechert (1897) for the solid part of the Earth, below the crust, "mantling" the material beneath or within. He thought it was made of rocky material, whereas the core was made of solid iron. But in 1906 the discovery of the seismic shadow revealed the presence of the liquid core (Oldham 1906).

The chief initiator of the "mobilist" view of the Earth's crust was the German meteorologist, geographer, geologist, astronomer, and explorer Alfred Wegener (1880–1930), who from 1924 held a chair in meteorology and geophysics at the University of Graz (Austria). He died tragically on an expedition on the Greenland ice cap.

Wegener arrived at the essentials of his theory in a few months in 1911 and made a public presentation of his ideas at the Frankfurt Geological Society early in 1912, publishing his paper later that year (Wegener 1912; Fritscher 2002). In this paper, he was reacting to problems in Suess's theory of Earth contraction as a cause of the formation of ocean basins and mountains. Wegener's ideas were later presented in *Die Entstehung der Kontinente und Ozeane* (*The Origins of Continents and Oceans*) (Wegener 1915, 4th English ed. 1971). Like others before him, he was struck by the complementary outlines of Africa's west coast and South America's east coast, and he remarked on the similarities of rocks and fossils on either side of the Atlantic—which suggested that the continents had formerly been connected and had somehow moved apart.

Wegener's title is interesting in itself. He was evidently responding to the *problems* he saw in the Suess paradigm. Isostasy was an important consideration. The ocean floors were known to be made mostly of dense "basic" rocks, such as basalt, containing *si*lica and *ma*gnesium compounds (called *sima* by Suess), while the continents were generally made of less dense rocks, such as the various sediments or granite, rich in *si*lica and *al*uminina (called *s[i]al* by Suess). So one might expect continents and mountains to stand higher than the ocean floors if their common substratum is plastic. On the other hand, why should the composition of the ocean floors and the continents be different if they were both formed by contraction and rupture of the same kind of material crust? Moreover, density factors would not, in themselves, account for the great folds and fractures in the mountain ranges or the enormous thicknesses of sediment in some of them, such as Hall had emphasized for the Appalachians. There was the further problem that Suess's huge "refuge" Gondwána—seemingly revealed by its ancient plants and animals in common—appeared to have broken up (by founderings of some of its parts). There were contradictions in geological theory.

A slow breakup and lateral movement of continents during geological time seemed to be required by evidences of Permian glaciation (e.g., scratched rock surfaces) in India, South Africa, and South Australia. One might expect such cold-climate phenomena to be associated with land situated near or around the South Pole. Glaciation in southern Australia and South Africa might not be too difficult to understand, but what was it doing in India? The rift valleys of Africa and such structures as the Red Sea seemed to suggest the incipient separation of continents, and rock types in common (e.g., schists and gneisses) in Canada, Greenland, parts of Scotland, and Scandinavia suggested a former connection, as did the Carboniferous coalfields of Europe and the northern United States at about the same latitude. On the other hand, the chemical composition of the volcanic rocks of the Pacific and the Atlantic regions differed significantly.

Considering such evidence, Wegener hypothesized that the present continents were joined during the Carboniferous in a single mass or "supercontinent" (Figure 8.6), which in later editions of his book he called *Pangea* (complemented by a "superocean": *Panthalassa*). He was able to offer some ideas about the

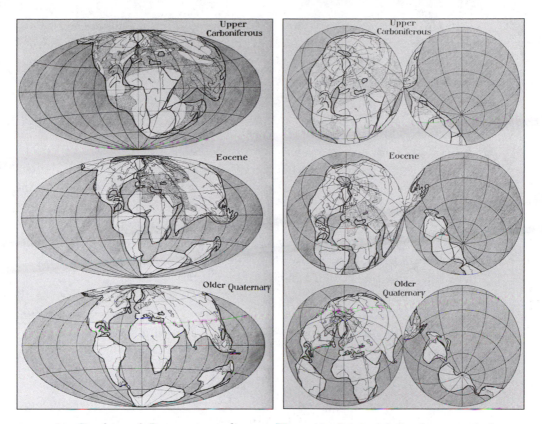

Figure 8.6: Breakup of Pangea, according to Wegener's *Origin of Continents and Oceans* (1966, 18−19). Image copyright History of Science Collections, University of Oklahoma Libraries.

order of the breakup and separation of formerly connected continental masses on the basis of paleontological evidence. For example, the Atlantic seemed to have split open from the south northward since the younger fossils in common were found on either side of the Atlantic in the northern parts, whereas further south one had to go back further in time to find fossil similarities.

Were this model correct, one might explain the formation of mountains such as those of the western sides of the Americas. If North and South America had moved away from Africa, then the forward edges of the moving continents might be crumpled, giving rise to great mountain ranges like the Andes. Moreover, volcanic activity in these ranges might be expected because of disruptions of the crust at the forward edges of moving continents. The lands around the Pacific seemed to be pressing forward into its ocean basin, with many chains of mountains and associated volcanoes. The Atlantic coastlines of, say, Ireland and Brazil were very different, without ranges parallel to the land, so that the Atlantic region seemed to be under tension. By its faunal relationships, India apparently had at one time been connected to Africa and Madagascar. So it could have separated, and, moving northward, there could have been a "collision" producing the Himalayas.

It will be seen that there were several components to the evidence mustered by Wegener. But many were unhappy with the idea of "continental drift." The idea of continents "plowing" through or across the ocean floors seemed inconceivable. What could be the forces producing such movements? Harold Jeffreys (1891–1989), professor of astronomy at Cambridge but working chiefly in geophysics, became one of Wegener's severest critics, though he acknowledged the strength of his paleontological arguments.

Wegener himself was unconvincing on the physical causes of "drift." Perhaps the movements were simply due to the Earth's rotation. Or there might be tidal forces within the Earth, caused by the gravitational attractions of the Sun and Moon. In his 1929 edition, he thought there might be convection currents within the *sima*, responsible for the continents' lateral movements. The idea of such convection currents had in fact been proposed by the Viennese geologist Otto Ampferer (1875–1947) in 1906, and it was also advanced by the British geologist Arthur Holmes (1890–1965) (1929).

Holmes was interested in radioactive minerals within the Earth and made important advances in determining the Earth's age by measuring decay rates for the recently isolated radioactive elements and estimating the global proportions of such elements and their different isotopes. In fact, he was the first to provide experimentally based arguments leading to the right order of magnitude for the Earth's age (see Lewis 2000). In 1929, Holmes published a paper with a diagram showing how he thought slow-moving convection currents within the Earth's viscous mantle might produce the effects of continental drift (Figure 8.7).

It will be seen from the upper picture that Holmes thought that magma might rise underneath a continent (where he thought the radiogenic heat might be trapped) and diverge, thereby causing a stretching and rifting of the crust above. This might explain, for example, the apparent incipient rifting apart of Africa at its Great Rift Valley. The lower picture shows a possible later situation in which new ocean basins had developed in the middle of the former continent.

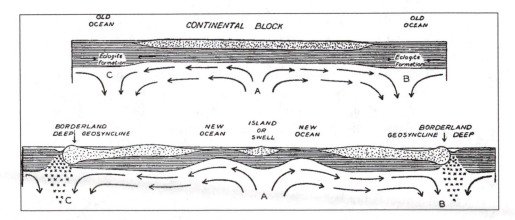

Figure 8.7: Convection currents within the Earth's mantle, producing continental drift (Holmes 1929, 579).

At the continental margins, two currents might meet from adjacent convection currents and descend together. This would be a region of high compression and attendant metamorphism. Minerals there would tend to change to high-pressure varieties (of lower volume and higher density), forming the "basic" rock called "eclogite," which would be carried into the Earth by the descending convection currents. The regions where the currents descended would be ones where we find ocean deeps today, lying seaward of island arcs or parallel to coastal mountain ranges, such as the Peru–Chile Trench to the west of South America.

The whole process was manifestly cyclic as far as the Earth's internal motions were concerned (and it provided a means for the dissipation of the heat constantly generated by radioactive minerals). The motions in the deep trenches accorded with the detection of deep-seated earthquakes occurring off the coast of places like Japan (an example of an island arc) (see p. 104). The forward-moving continental margin would be compatible with its strata being severely distorted and mountains thrust up. The continental crust would become thickened, and the mountain roots, unable to sink along with the dense eclogite, would fuse and give rise to the volcanic activity seen around the Pacific, with the rock type "andesite" (produced in the Andes!) being formed in abundance in active volcanoes.

Holmes also suggested that parts of the material of the lower surface of the continental crust would be stripped away as the convection current passed under it, being replaced by denser material of the substratum. So the region would subside where this occurred, and thus a geosyncline would be formed where sediment could accumulate as envisaged by Hall and thus cause further subsidence.

Holmes (1929, 582–83) continued,

> Sooner or later a time will come when the buried crustal material beneath the infilled geosyncline will become hotter and weaker than the material behind and in front. Moreover, the grip of the underlying current will have weakened as a result of the higher temperature and lowered viscosity. At this stage thickening of the geosynclinal belt due to compression will become more effective than thinning by magmatic corrosion; mountain building will have set in [Figure 8.8]. . . .
>
> Meanwhile, the mountain roots will set up an accelerated transport of heat towards the continental edge, and the currents will thus move out far beyond the edge before beginning to descend. The borderland or mountain arc will then no longer be in the original zone where the horizontal component of rock-flowage is retarded against the belt of eclogite formation and deeps. Its lower levels will be stretched and thinned, and the area will subside as the interior geosyncline had previously done.

Thus, a *succession* of geosynclines was envisaged—or cycles, we might say. It should be remarked how Holmes sought to produce a theory that synthesized his knowledge of the generation of heat by radioactive minerals, convection theory, and geosyncline theory to provide a theory to explain the elevation of mountains from the sediments in geosynclines and to account for the data about the breakup of continents that Wegener had contemplated.

It also avoided the "plowing" difficulty. Holmes went on to describe other types of geosynclines, such as those formed where two continental blocks separate (Figure 8.7) or where two forelands approach one another. Later writers, such as the German Hans Stille and the American Marshall Kay, developed a whole "menagerie" of different kinds of geosynclines—which we have no desire to visit here!

Despite Holmes's eminence and the fact that he made his ideas known in one of the most successful geological textbooks ever published (Holmes 1944), his "mobilist" theory of 1929 did not catch on. There were, to be sure, other supporters of mobilism, such as the Swiss geologist Émile Argand (1879–1940), who converted to Wegener's views as early as 1915. He thought that the Himalayas were produced by a collision between India (drifting away from Africa) and the larger continent of Eurasia, which had closed Suess's Tethys Sea in the process (Argand 1924, 1977). In the 1930s, the South African geologist Alexander Du Toit (1878–1948) also accepted the *fact* of drifting at least and produced a comprehensive account of the evidence (structural, topographic, paleontological, climatic, and petrologic) in favor of the breakup and lateral motions of continents. He also gave a valuable synopsis of the ideas of geologists who had favored or opposed mobilism. Today it may seem that the weight of evidence lay on the side of mobilism, which "should" have been accepted long before the 1960s. But the influence of geophysics was strong. The forces that might be required to move continents across the globe seemed inconceivably great, and the idea of transient east—west land bridges (analogous to the Panama isthmus) to account for the paleontological evidences was mostly preferred. Du Toit proposed that there were two great supercontinents in the Paleozoic—which he called "Laurasia" and "Gondwana"—that began to break up in the latter part of the Mesozoic. In this respect, his thinking was different from Wegener's idea of "Pangaea."

Du Toit (and others) had the idea of a succession of periods of folding and mountain formation, called orogenies or "diastrophisms": the Caledonian, Hercynian, and Alpine Earth movements being the main ones (but Du Toit tabulated others besides). In this sense, there was an element of cyclicity to his theory, though Du Toit questioned whether convection currents could explain everything. He concluded,

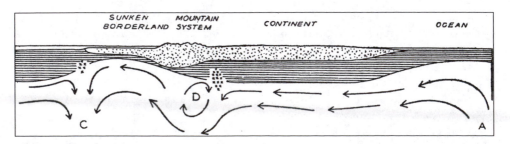

Figure 8.8: Development of geosynclines and the formation of mountains (Holmes 1929, 582). Cf. Figure 8.7.

In our interpretation, every revolution has involved a certain amount of "drift" and such movement must hence be regarded as a normal consequence of the orogenic cycle. The *Energy* therefore must be derived from some internal source that is itself cyclic, and of such the only outstanding one is that provided by the Earth's radioactive content. *Drifting must indeed be viewed as a function of Radioactivity.* (Du Toit 1937, 322; emphasis in the original)

That is, the gradual buildup and release of radiogenic heat could lead to great cycles of mountain building (diastrophism).

In his *Physical Geology* (1944), Holmes offered a somewhat revised depiction of his ideas on convection currents (Figure 8.9). He suggested the formation of "ocean deeps" (geosynclines) and adjacent mountains in areas of compression. But the ascending convection current now appeared to be able to give rise to a new ocean into which fragments of the old continent might founder. The process involved cycles of mountain building when "the convective circulations became unusually powerful and well organised" (506).

Holmes adduced much new information from geomagnetism (see Chapter 12) in his second edition (1965). When rocks are formed, they become slightly magnetized (from the Earth's field), with their magnetic axes directed along the magnetic meridian that prevails at the time of their formation. This orientation is maintained, even if the rocks are subsequently moved around (as when continents change their location on the Earth's surface by some process of "drift"). So, if rocks' "remanent" magnetism is ascertained and their ages are known, one can tell approximately where the Earth's magnetic poles were at the time of their formation, and the evidence indicated that the poles had shifted substantially through geological time. Indeed, the paths of the shifting magnetic poles could be traced on a modern globe. Such work was conducted particularly in Britain by Paul Blackett (1961) and Keith Runcorn

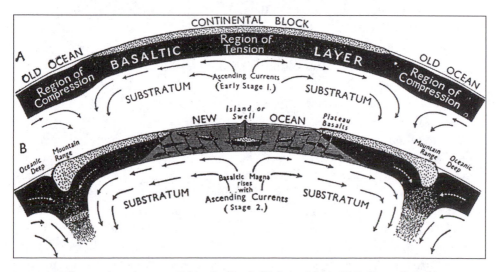

Figure 8.9: Convection currents within the Earth (Holmes 1944, 506). Reproduced by permission of Taylor & Francis Ltd.

(1962) and by workers at the Australian National University, Canberra. It provided powerful evidence for continental drift and came to be seen as such after about 1956. (On geomagnetism, see Chapter 12.)

The main objections to "drift" theory came from the English-speaking world and especially the United States (Oreskes 1999). The events that led to the overthrow of the "fixist" views, which went back at least as far as Dana, and their replacement by plate tectonics has been told many times (e.g., Glen 1982; Hallam 1973; Le Grand 1988; Marvin 1973; Menard 1986; Oliver 1996; Oreskes and Le Grand 2001) and will be only sketched here.

The main things were the discovery by underwater survey of the system of mid-oceanic ridges. These had been known in a general way since the work of the *Challenger* expedition, but the details became much clearer after World War II through American survey work (much of it for military purposes). It became evident that the ridges tended to have the same form as their adjacent coasts: for example, the one in the south Atlantic had the same shape as the western coast of Africa and the eastern coast of South America. Further, the ridges were found to have narrow "rift valleys" running along their submarine summits.

Then a pattern of deep-seated earthquake foci was found, near the deep-sea trenches, marking out steeply inclined planes that seemingly passed deep into the Earth (e.g., off the coast of Japan). The Japanese seismologist Kiyoo Wadati (1902–1995) had earlier studied earthquake patterns around Japan and published a series of papers on the topic in the 1920s and 1930s (Suzuki 2001). His evidence suggested that earthquakes could occur at greater depths than previously supposed, and in his classic paper (Wadati 1935), he showed that when the focal depths of intermediate and deep earthquakes in his region were contoured, they revealed a pattern of inclined surfaces, dipping away from the oceanic trenches. And the volcanic zones of the Japanese archipelago coincided with the lines of equal depth for intermediate earthquakes. He thought that the dipping surfaces might indicate lines of weakness in the Earth's crust and might be linked to Wegener's theory.

Wadati's ideas were "rediscovered" in the 1950s by Hugo Benioff (1899–1968) (1954) of the California Institute of Technology. By studying the records of many earthquakes, he, like Wadati, found evidence of steeply dipping fault planes. Those found in oceanic regions appeared to extend down to about 700 kilometers, while others at continental—ocean margins dipped down toward and under the continents, again to about 700 kilometers but changing their inclination at a depth of about 300 kilometers. At the time, this evidence did not compel belief in continental drift (or plate tectonics). But subsequently, the great faults (now called Wadati–Benioff zones) became an important part of the evidence supporting mobilism, for the zones could be interpreted as areas where crustal material was plunging toward the Earth's interior as part of great convection cycles within the mantle. Benioff suggested that the heat generated by the friction in such zones could cause melting and hence the eruption of volcanoes in island arcs or coastal ranges, where seismicity is intense.

Clinchers for the acceptance of the theory that the processes of forming mountains and ocean basins depended on cyclic movements of semiplastic rock in the Earth's mantle came from geomagnetic evidence for "seafloor spreading" and the idea of a new kind of faulting on the ocean floors that made such spreading possible. Harry Hess (1906–1969), appointed to Princeton in 1932, served in the U.S. Navy in the Pacific theater during World War II and took the opportunity to conduct depth surveys by echo sounding. His investigations revealed the existence of flat-topped submarine mountains (resembling volcanoes with their tops cut off; see p. 83), which he later called "guyots" after the Swiss American geologist Arnold Guyot (1807–1884), Princeton's first professor of geology. Hess was also a devotee of Holmes's ideas, and when the mid-oceanic ridges were shown to have rift valleys running along their axes, he proposed the idea of convection currents in the mantle, with basaltic magma rising to the surface at the ridges and then spewing "sideways," eventually to plunge down again into the mantle at what were later called "subduction zones" (which could be correlated with the Wadati–Benioff fault zones). This model was proposed orally in 1959, circulated as a preprint the following year, and published two years later (Hess 1962). The continents were not plowing through the basaltic ocean floors but riding on the mantle material that had risen from below and was moving away laterally from the mid-ocean ridges.

Similar ideas were proposed independently by Robert Dietz (1914–1995) (1961), who proposed the term "seafloor spreading." As a marine geologist trained at the Scripps Institute of Oceanography in San Diego, California, working with Francis Shepard (see p. 134), Dietz was invited to organize a sea-floor studies group at the U.S. Navy Electronics Laboratory at San Diego. From that base, he became interested in the sediments of the submarine shelves and their submarine canyons, which channeled sediments into the ocean deeps (see p. 106-107), and he worked in a group mapping the canyon profiles. A year in Japan in 1953 led him to study the Pacific region. The chain of "guyots" that extended northwest from Hawaii, with the depth of their tops gradually increasing with their distance from Hawaii, suggested that old volcanoes were being carried north by some kind of "conveyor belt." This became part of his evidence in favor of seafloor spreading.

But it was geomagnetic evidence that yielded the most persuasive evidence leading to the collapse of the geosyncline paradigm. Not only did paleomagnetic investigations on land suggest that the Earth's magnetic poles had formerly occupied different positions, relative to the present landmasses (Runcern, 1963), but examination of the magnetic characteristics of the basalts of the ocean floors revealed that there were alternating "stripes" of different magnetic orientation (Figure 8.10). Such surveys were conducted by U.S. oceanographic survey vessels in the 1950s, chiefly from the Scripps Institute, the results being obtained by towing magnetometers behind the ships. It can be seen from the figure that the magnetic "stripes" are apparently offset along certain lines by what appear to be fault lines.

In 1962, vessels from Lamont and Woods Hole laboratories, on the East Coast of the United States (see p. 105), found that the Mid-Atlantic Ridge was offset in some places by fracture zones, and similar fractures had also been found in the Pacific seafloor. Then two scientists from the geophysics unit at Cambridge University, Frederick Vine and Drummond Matthews (1963), provided a synthesis of the evidence by a grand hypothesis: lavas were generated along the lines of oceanic ridges and moved *laterally,* constantly generating new seafloor. The youngest basalts would be nearest the ridges and the oldest furthest from them. As the basalts were of different ages according to their distance from the ridges, they could—assuming that the direction of the Earth's magnetic field reversed from time to time—have alternating north and south polarities, as illustrated in Figure 8.10. (Essentially the same hypothesis had been proposed earlier by Lawrence Morley to the Royal Society of Canada in June 1963, but his paper was refused by both *Nature* and the *Journal of Geophysical Research,* and his ideas were published subsequently [Morley and Larochelle 1964].)

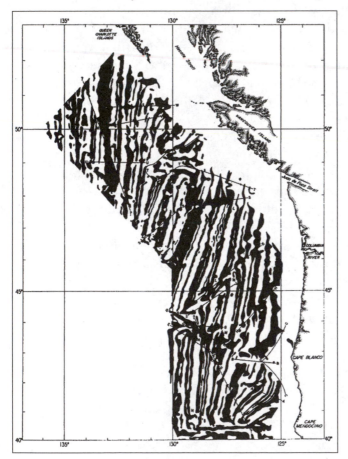

So there would be a "motor" for continental drift if magma could rise from the Earth's interior, move laterally, and then sink back again at subduction zones, manifested by the Wadati–Benioff zones and the deep oceanic trenches. There would be gigantic subterranean convection cycles, driving the process of "drift," and the surface movements on the ocean floors would be manifested by the magnetic reversals, which could themselves be thought of as two-phase cycles (or as a kind of magnetic "tape recording"). The fracture zones represented adjustments of the lavas, delivered to the surface from the Earth's interior and moving on its curved surface. The descending chain of seamounts, such as Dietz investigated near Hawaii, indicated that they were being gradually moved from their center of origin (Hawaii) by a slowly moving and descending ocean floor.

Figure 8.10: Magnetization of the rocks of the ocean floor off the coast of western Canada (Mason and Raff 1961, 1268). The black and white areas indicate the presence of seafloor basalts with opposed directions of magnetization, acquired at the time of their cooling and solidification. Reproduced by courtesy of the Geological Society of America.

Not even the visual imagery of the "magnetic stripes" led to general conversion to mobilist theory. But the advocacy of the Canadian geologist J. Tuzo Wilson (1908−1993) of Toronto University finally turned the tide in favor of "mobilism" and what came to be called "plate tectonics." The idea of the movements of separate parts of the lithosphere (plates) over the Earth's interior was partly generated in Wilson's mind by his flying over the Hawaiian Islands and remarking that they appeared to be larger and more volcanically active to the southeast and smaller and less active or inactive to the northwest. This suggested that the islands' volcanoes were produced by the crust moving northwestward over some subterranean "hot spot." The idea was later confirmed by radiometric age determinations of the islands' rocks, which showed that the more southerly ones were indeed younger and the more northerly older. Wilson (1965) also proposed the idea of so-called transform faults. These occurred where two plate boundaries or contacts "rubbed past" each other in the course of the plates' motions. These faults were such that they suddenly terminated, or "changed their form," often at mid-oceanic ridges: they allowed the plate motions to be accommodated on the Earth's spherical surface.

The plate tectonic theory, gradually assembled by the previously mentioned workers (and many others), was consonant with but different from earlier mobilism. It accounted for the phenomena that Wegener had thought so relevant to his hypothesis (the "fit" of continents or the distributions of animals and plants on the different continents). And the idea of the processes being driven by convection currents within the Earth's mantle, themselves driven by the heat liberated by the decay of radioactive elements, was more attractive than the continents "plowing" though the ocean floors or being driven by forces generated by the Earth's rotation.

But did it imply that the Earth had a one-way, noncyclic history—for presumably Wegener's Pangea could break up only once? Well, by no means, for there is no reason in principle why the separate continents might not also recombine after many vicissitudes. This process could result from the various openings and closings of ocean basins in the so-called Wilson cycle, named thus in honor of Tuzo Wilson, who first suggested that the Atlantic Ocean might have closed and subsequently reopened (Wilson 1966) and is reported by Kevin Burke to have suggested to the American Philosophical Society at a meeting in 1968 that it would be "appropriate to interpret earth history in terms of the life cycles of the opening and closing of the ocean basins" (http://geowords.com/histbooksnetscape/j23.htm).

Imagine a large continental mass, bordered by oceans—perhaps like Australia. Suppose heat is trapped underneath the continent and magma rises under it from a subterranean source ("plume") or a chain of such plumes. This may cause rifting or splitting of the continent into two parts (A, B), as is apparently happening today in Africa. Basaltic magma erupting along the rift may form a new ocean floor there, and sediments may become deposited in the basin. With cooling and subsidence of the lands of the two new continental margins, the water in the basin may begin to encroach on the land. But as the continents diverge, there will be a continuing mid-oceanic ridge acting as a source of new magma and producing a new ocean floor.

The receding continents will gradually cool and sink, even below sea level, but before that a great wedge of sediments (some, like coral reefs, being the products of living organisms) will have built up at the sides of the basin adjacent to the receding continent (as in the coastal waters off Virginia in the United States today). The new oceanic crust may, however, eventually fracture and subduction begin somewhere in the basin, and its material passes down into the mantle as the convergence phase of a Wilson cycle begins.

There can be two kinds of subduction zone. If it occurs *within an ocean basin,* an island arc forms (e.g., the Aleutian Islands) as a result of volcanic activity generated by the carrying down of wet sediments into the Earth's hot interior. Alternatively, subduction may occur *at the edge of a continent* (say, *B,* as in western South America). Large masses of granite (batholiths) may be intruded as well as volcanoes rising and erupting. The area on the subduction trench side of the line of volcanoes is called the "fore-arc," where compression occurs; the other side is the "back-arc," an area that experiences tension. In an island-arc situation, large masses of sediments can be formed from the erosion and weathering of the volcanic masses, and these spill onto the ocean floors. For an Andean situation, sediments can likewise be washed into the fore-arc's trenches, but on the other side of the rising mountain range, they may have to be carried long distances to the sea by rivers, as with the Amazon.

Because of the convergence situation between the two continents (*A, B*) and the subduction, the area between the continents will be squeezed so that it eventually becomes no more than a remnant ocean, and its floor is destroyed. The resulting junction between the old continent (*A*) and the volcanic arc rocks and associated sediments, plus the remains of the intervening ocean-floor basalts, forms a "suture zone," consisting of sheared and altered rock. Usually the former island-arc material or the old continent (*B*) plus its volcanic materials is forced over the old continent (*A*), the two parts being called the foreland and hinterland, respectively—which recalls the terminology of Suess (p. 87). The overriding hinterland rocks form mountain ranges, while the foreland rocks buckle and form basins, which then fill with sediments. In time, the new "compound" continent is reduced to a peneplain, approximately at sea level. The net result of all this is that the original continental core (*A,* the foreland) has material (from *B*) "sutured" onto it, so it has "grown" (as Dana would have approved!). Suture zones can be recognized in the rocks. For example, there is one between northern England and southern Scotland.

The foregoing processes could be repeated by the formation of another island arc and its addition to the newly formed continent, or the other portion of the originally fractured continent might be brought into play. That is, by lateral forces, the newly enlarged continent might approach the other half of the originally hypothesized single continent, with the ocean between them closing and subduction occurring toward the second continental mass. (Alternatively, the subduction could proceed at the other side of the intervening ocean, adjacent to the previously sutured continent.) This would elevate a large marginal mountain range, such as the mountains of Washington and Oregon or the Andes. But another convection current under its hinterland could produce tension and subsidence and the formation of a back-arc sediment basin in that area. This is, then, the product of a continent—continent collision, such as is occurring today in the Himalayas. Eventually, all will be eroded to a new large stable continental block, but with a complex internal structure, showing the relics of former continent—continent collisions.

The net result of the sequence of the events outlined in the preceding paragraphs takes one, via tectonic processes (and theory!), from a single stable continental block (a "craton" or a "platform") to another one of more complex structure in a process involving the opening and closing of oceans. This is the Wilson cycle. It should be noted that it involves the formation of deep-sea trenches and their filling with sediment and the accretion of new materials onto an old stable block. In these respects, it gives an alternative to geosyncline theory but in terms that invoke large-scale physical processes within the Earth (mantle convection currents and hot plume material rising from the fluid core). Since the processes may in principle be repeated many times, a single continental mass may have many constituent components (terranes) stitched together, so to speak. So the growth of continents, as envisaged by Dana, is accounted for by the Wilson cycle. (For depictions of the cycle, see, http://csmres.jmu.edu/geollab/Fichter/Wilson/Wilson.html or http://loki.stockton.edu/~epsteinc/wilnote.htm.)

It's perhaps worth noting that the condition of the globe at the time of Pangea, with only one supercontinent, may be compared with its configuration as envisaged back in the Middle Ages by the likes of Buridan (see p. 17) or later by Leonardo da Vinci. But the resemblance is coincidental. On the other hand, one may fairly ask whether supercontinents may have been formed repeatedly on the Earth over immense periods of time, so that their repeated formation may be regarded as constituting some even greater cycle. This idea has been advanced by, for example, Brendan Murphy of St. Francis Xavier University, Nova Scotia, and Damian Nance of Ohio University (Murphy and Nance 1991, 1992), in what they (and others) have called the "continental supercycle." They envisaged the periodic division and dispersal of continents and then their "reeling in" by subduction processes, so that a supercontinent forms perhaps once every 500 million years. That means there will only have been one supercontinent between the Cambrian and the present (i.e., during the so-called Phanerozoic; cf. Greek *phaneros* = "visible" or "evident"; *zoe* = life). Yet that is little more than one-tenth of the Earth's total age. So, by Murphy and Nance's thinking, the supercycles must have started back in the Precambrian.

The oldest clearly recognized supercontinent has been called Rodinia (from the Russian *rodina*, "homeland," or *rodit*, "to give birth"), a term that came into use in the 1990s (McMenamin and McMenamin 1990, 95). Rodinia is thought to have assembled about a billion years ago. Suggestions have also been made for earlier supercontinents, the oldest (rather speculative one) being named "Ur" (= "original"), which may have formed about 3 billion years ago. "Arctica" has been suggested as forming at about 2.5 billion years and "Baltica" and "Atlantica" each at about 2 billion. "Nena" (supposedly made up of parts of what are now northern Europe and North America) formed at about 1.6 billion from Arctica and Baltica. At about 1.1 billion, Nena, along with Atlantica and Ur, became part of Rodinia, which broke up at about 0.7 billion into East and West Gondwána and Laurasia, all of which eventually joined up again to form Wegener's Pangea. These ideas are based on John Rogers (1996), but other "scenarios" have been offered, and the dates in the current literature vary somewhat. By Rogers's account, there have been only three supercontinents—Ur, Rodinia, and Pangea—but Nena is sometimes represented as a supercontinent. For a series of illustrations of the supposed "dance of the continents," see Dutch (1999), but no particular scenario currently receives general acceptance. The hypothesized earlier continental masses are suggested on the basis of different "components" of Pangea apparently being deformed at the same time and the radiometrically determined ages of the different parts. The way such grand cycles may be linked to geochemical cycles will be considered in Chapter 13.

I conclude this chapter by turning from "megascopic" cycles to "microscopic" ones, revealed in the sediments of sedimentary basins.

For half a century or more, there has been great interest in what are called "turbidity currents" (mud-laden bodies of water flowing downhill but underwater), the sediments produced by them being called "turbidites." The former term was probably first introduced by Douglas Johnson (1878—1944) (1939, 27), a geomorphologist at Columbia University, though the notion itself goes back well before that. The Swiss geomorphologist François Alphonse Forel (1841—1912) (1885) described the *underwater* channels cut into the river deltas by the Rhine (Lake Constance) or the Rhone (Lake Geneva) with underwater levees on each side. He proposed that the channels were cut by the dense low-temperature, sediment-carrying, fast-flowing river waters, while the levees were produced by deposition of silt in the quieter waters at the sides of the channels. The denser waters could cut channels in the soft lake sediments. Such currents thus became known as "density currents."

In the 1920s and 1930s, attention turned more to the consideration of the channels (or even canyons) that depth studies revealed on the sides of the submarine sediment shelves that are found around landmasses capable of delivering large quantities of sediment. Were these canyons formed by slumps (perhaps triggered by earthquakes), or were they produced by the action of turbidity (density) currents, analogous to those investigated by Forel? Or could they simply be drowned valleys, like, say, Sydney Harbor? The first hypothesis was supported by the fact that submarine cables were known to have got "chewed up" by earthquakes, such as one off the coast of Newfoundland in 1929. But Johnson (1939, 12) doubted whether earthquake-induced slumps were responsible for the telegraphy disruptions. Direct depth soundings certainly suggested changes in submarine topography before and after earthquakes, but the data had not been collected with the intention of testing the hypothesis, and thus not too much reliance could be placed on them. So while Johnson countenanced occasional submarine mud slides, he doubted that they produced canyons.

Subsequently, geologists came to the view that turbidity currents could have considerable erosive power and could move large boulders as well as gouging channels in the soft sediments of offshore shelves. Or the canyons might be caused by slides, just as hillsides may sometimes slump, whether or not they are triggered by earthquakes. The sediments may move slowly or at high speeds, up to (or even more than) 100 kilometers per hour. And they can spread sediment for hundreds of kilometers over the oceans' abyssal plains. Their formation can be replicated experimentally by allowing sudden discharges of sediment into a tank of water by opening a "gate" (e.g., Kuenen 1948; Kuenen and Migliorini 1950). When this is done, one sees a kind of lobed front of sediment rushing into the tank, and as this settles, the coarser and heavier particles settle first, producing "graded bedding." Such work on turbidites by the Dutch sedimentologist Phillip Kuenen (1902–1976) was used to make sense of the so-called flysch deposits found in the Alpine areas of Europe, which had previously been somewhat mysterious in origin. (*Flysch* was a provincial Swiss term used to refer to a series of dark slates, marls, and sandstones of Tertiary age, exhibiting rhythmically interbedded coarser materials.)

What interests us here is that such sedimentation can exhibit cyclicity if an episode of sediment deposition occurs repeatedly—that is, by repeated slumpings on a continental shelf. Such repetitions were studied by a Dutch sedimentologist and student of Kuenen, Arnold Bouma (b. 1932), in his fieldwork as a young geologist, examining turbidites in the area of Peira Cava in

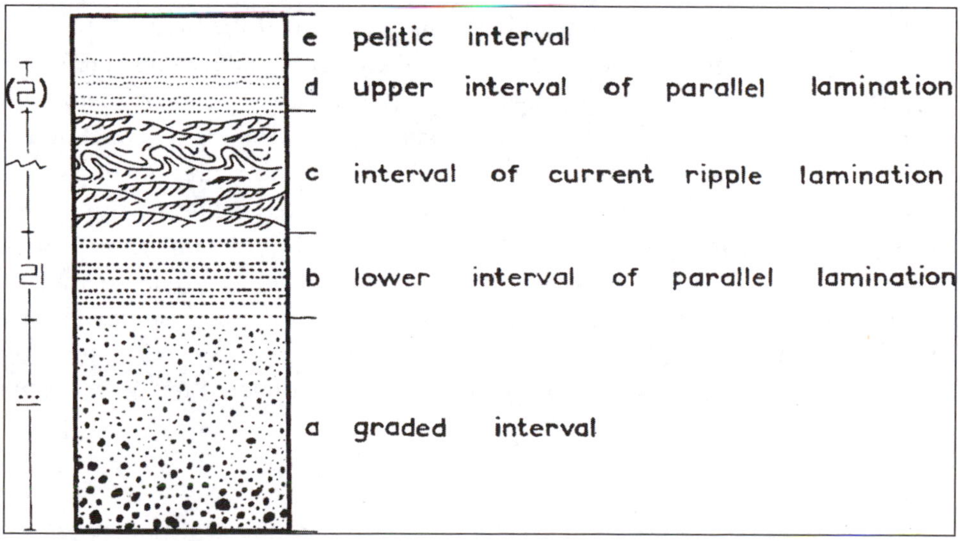

Figure 8.11: The "ideal" Bouma sequence, as depicted by Bouma (1962, 49). Reproduced by permission of Professor A. H. Bouma. Image redrawn from original by Ricochet Productions.

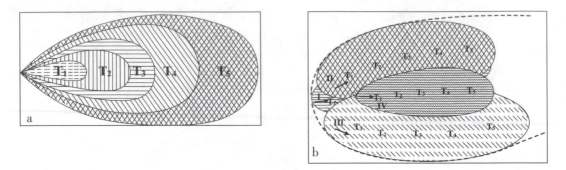

Figure 8.12: Hypothetical form of the deposition cone of a turbidity current and hypothetical filling of a basin by turbidites.

(a) Hypothetical form of the deposition cone of a turbidity current, based on Bouma (1962, 99). The sequences may be more or less complete and were named by Bouma as T_1, T_2, T_3, T_4, and T_5. $T_1 = A-E$; $T_2 = B-E$; $T_3 = C-E$; $T_4 = D-E$; $T_5 = E$.

(b) Hypothetical filling of a basin by turbidites, based on Bouma (1962, 99).

$\rightarrow$ = Current Direction. Images redrawn from Bouma's originals by Ricochet Productions.

southeastern France. (For photographs of the sediments and their structures in this area, see Locke 2000.) Bouma observed the following repetitions:

E (top): pelitic material (slowly accumulated mud)

D unit: deposition of parallel-laminated silts and muds

C unit: deposition of current-rippled layers

B unit: deposition of parallel-laminated sands

A (bottom): deposition of massive sands or conglomerates (generally graded in grain size, with coarse at bottom and fine at top) with a clear-cut base

These are represented in Figure 8.11. The whole came to be known as a "Bouma sequence," a term common in the turbidite literature. The way in which series of such sequences may be formed is shown in Figures 8.12a and 8.12b.

So we have a kind of sedimentary cyclicity. I have gone into this point in a little detail, partly because it concerns what goes on in sediment basins and also because it takes us back in a remarkable way to the eighteenth century. If readers will refer to Figure 5.5, they will see Hutton's representation of the Jedburgh unconformity with what we now call Old Red Sandstone (Devonian) lying atop vertical Silurian greywackes, which are in fact turbidites. Inspection of the figure reveals some significant structures in the vertical sediments, which very likely were representations of Bouma-type structures—not any old type of decorative shading!

We shall further consider the question of cyclic sedimentation recurs in the following three chapters.

9

THE EARTH, THE SOLAR SYSTEM, CYCLES, AND GLACIATION

One of the problems—and achievements—of geology has been to understand the causes of glaciation and to integrate glacial theory with understanding of the general history of the Earth and the workings of the solar system.

The idea of a greater former extent of glaciers when mountains were higher before being reduced by erosion was suggested by Hutton (1795, 2:218), and the notion of a former extension of glaciers was proposed more explicitly by the Swiss naturalist Louis Agassiz (1807–1873) (professor at Neuchâtel) in 1837. The past extension of the Swiss glaciers is easy to recognize. The modern glaciers can be seen descending their valleys, and old moraines can be identified well below their present terminal moraines. Moreover, large rounded boulders of identifiable rocks from the main Alpine ranges can be found on the slopes of the Jura Mountains to the north of Lake Neuchâtel on the opposite side of the valley from the Alps proper.

To explain such observations, Agassiz assumed that the world had formerly been colder and that glaciers once extended all over the Alpine region, depositing cargoes of boulders (ripped from the hills by the moving ice) when the ice subsequently retreated. The retreating glaciers also deposited a mix of boulders, clay, sands, shells, or anything else that had stood in their path. This was Buckland's "diluvium." Today it is called boulder clay or "till," and if consolidated into a rock, it is called "tillite." Agassiz also remarked on the evidence for former extended glaciation in the scratches on smooth rock surfaces. Such scratches could have been gouged by rock fragments frozen into the bottom of glaciers, and the orientation of the striations revealed the direction in which the glaciers had traveled.

While the phenomenology of glaciation was quite easy to comprehend, its causes were another matter. Agassiz (a specialist in modern and fossil fishes) was a disciple of Cuvier (see p. 64), whose catastrophist ideas he imbibed.

Having developed the idea of the former widespread glaciation of the Alpine regions, Agassiz adumbrated the notion of a former *Eiszeit* (Ice Age) in a lecture delivered to the *Société Helvétique* in Neuchâtel in 1837, later called the *Discours de Neuchâtel* (Agassiz 1837, 1838).

Like many of his contemporaries, Agassiz accepted the idea of a cooling Earth, but to explain the idea of an *Eiszeit,* he assumed that the cooling occurred in a curious fashion. His suggestion was that it proceeded stepwise (Figure 9.1). *If* this were the mode of cooling, *then* today's temperature might be somewhat higher than in the fairly recent past, even if there was an overall cooling. Moreover, Agassiz thought that each dip in the imagined cooling "curve" corresponded to a Cuvierian catastrophe, after which there was a re-creation of new organisms. Thus, Agassiz's theory was intended to deal with general features of the biostratigraphic column, interpreted through the lens of Cuvier's geotheory.

Agassiz wrote a detailed account of his theory in 1840 and traveled to Britain that year to attend the British Association meeting in Glasgow. Afterward, he went with Buckland to look for Scottish evidence in favor of the theory, and their efforts appeared to be successful. So Buckland renounced diluvialism and became an advocate of the glacial theory. Lyell was temporarily converted but later reverted to his old theory of climatic cycles (see p. 65): Agassiz's ideas were too catastrophist for his taste! As we have seen, Lyell suggested that the erratic boulders could have been transmitted by floating icebergs, which was by no means impossible and is still accepted as a means of depositing "drop stones."

In the next few years, there were various modifications of the glacial theory, notably that the mixtures of alternating layers of sand and gravel and boulder clay indicated the occurrence of several glacial episodes. It was also suggested that glaciation might be accompanied by the encroachment of the seas on the land, to explain the intermingling of the sediments, in the previously mentioned "glacial submergence" theory. (This was implausible, as lowering of the land surface would not be expected to accompany a deteriorating climate, and if much of the world's water became locked up in ice, one wouldn't expect sea levels to rise.) The problem of lakes also became controversial. The glacial theorists contended that hollows could have been scooped out by moving glaciers. Their opponents disagreed, but the arguments of the British Survey officer Andrew Ramsay (1814–1891) (1862) on the question of glacial lakes eventually proved persuasive. The cold-water marine shells found near the

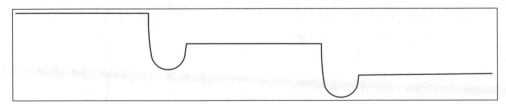

Figure 9.1: Global changes in temperature over time (Agassiz 1837). Image redrawn by Ricochet Productions.

top of a small mountain in northwestern Wales led the ice theorists to suggest that they had been scooped up from the area of the Irish Sea and deposited about 1,000 feet above the present sea level by a large mass of ice moving over Wales.

But what was the cause of it all? Agassiz's cooling "curve" was no explanation!

Changes in the heat received from the Sun was a possibility, either because of changes in the energy *emitted* by the Sun or the annual amount of radiation *received* on the Earth due to alterations in its orbit. The latter could be due to changes in the inclination of its axis relative to the plane defined by the Sun and the Earth's orbit, changes in the orbit's shape, or changes in the orientation of the Earth's axis relative to the fixed stars, which produce the "precession of the equinoxes" (Figure 9.2).

The Earth's axis makes a slow precessional rotation (clockwise if viewed from above the North Pole) with a period of about 26,000 years. But its elliptical orbit also rotates (counterclockwise from the same viewpoint), and the combination of these two effects yields the observed precession of the equinoxes, which has a period of about 22,000 years. Such matters were sorted out by the French mathematician and astronomer Jean d'Alembert (1717–1783).

In 1830, John Herschel suggested that astronomical causes might cause climate changes and produce significant geological effects (Herschel 1826–1833). He thought that a more elliptical orbit (with a smaller minor axis for the ellipse) would increase the annual solar radiation received by the Earth. But this idea was not taken further by him (and the effect was later regarded as insignificant).

Then, drawing on d'Alembert's ideas, the French mathematician Joseph Alphonse Adhémar (1797–1862) (1842–1844) pointed out that in half a cycle of precession, one of the Earth's hemispheres would have a longer and the other a shorter winter and vice versa in the other half of the cycle. Moreover, he suggested, the hemisphere with the longer winter would have a glacial period. Thus, there would not be *global* glacial epochs, as Agassiz envisaged, but alternations of glaciation between one hemisphere and the other.

But the influential Alexander von Humboldt (1769–1859) rightly pointed out that regardless of changes in the ellipticity of the Earth's orbit, the *total* amount of heat received by each hemisphere in a year would be the same— a longer winter being compensated by a shorter but hotter summer. So an astronomical theory of glaciation was put aside for a while.

It was, however, revived in 1864 by a remarkable Scottish autodidact, James Croll (1821–1876). Croll came from a working-class family and started life as a carpenter, without much success in that line of work or in his later attempted business ventures. Eventually, he became a caretaker at Anderson's College, Glasgow, where he used its library, came across Adhémar's work, and began to develop his own ideas on the topic. (Croll was already writing on philosophical and theological matters.) He succeeded in getting his ideas printed in the *Philosophical Magazine* (Croll 1864, 1866, 1867) and began to correspond with some of the leading British scientists of the day. His ideas impressed Archibald Geikie, the new head of the Scottish Branch of the Geological Survey, and Croll was appointed to its staff in 1867. He was unlike the other surveyors, doing no formal mapping. But he was regarded as a genius by his colleagues and was allowed to do indoor work and publish his theoretical ideas. This work culminated in a famous book, *Climate and Time* (Croll 1875). Although he did not do mapping, Croll did study the physical manifestations of glaciation in the field.

Archibald Geikie's brother James (1839–1915) was also on the Scottish Survey staff and specialized in relatively recent geology and glaciation. By the 1870s, it was becoming clearer that there had been more than one episode of glaciation in Europe in the recent geological past (see p. 110), and this needed to be taken into account by any glacial theory. Moreover, evidence was emerging of glaciations way back in the stratigraphic record (e.g., in the Permian), which also needed to be accounted for. Croll's theory was intended to deal with such matters and also the fact that other geological epochs, such as the Cretaceous, betokened warm conditions.

Croll discounted Herschel's suggestion that changes in the ellipticity of the Earth's orbit could significantly alter the heat received by the Earth from the Sun each year. Nevertheless, changes in the ellipticity (arising from the changing gravitational attractions of the Sun, the Moon, and the various planets) were at the heart of Croll's theory. Croll (1866) calculated the eccentricity of the Earth's elliptical orbit and the longitude of its perihelion (see the discussion later in this chapter) at 100,000-year intervals—1 million years into the past and 1 million into the future—using formulas developed by the French astronomer Urbain Leverrier (1811–1877), that he found in a book by Louis-Benjamin Francoeur (1773–1849) (1843). Croll's results were later presented graphically in *Climate and Time* (chart facing p. 313). The chart showed cycles in the eccentricity of the ellipse of the Earth's orbit, with clusters of peaks occurring every few hundred thousand years. So if climate were somehow related to orbital eccentricity and precession, then one might expect to see several glacials and interglacials bunched together and then intervening periods of milder climate when the orbit was less eccentric.

One had to consider two interacting cycles: the "waves" of increasing and decreasing ellipticity and the precession movement of the Earth's axis of rotation, which could swing the equinoctial points around the heavens, bringing first the northern hemisphere and then the southern hemisphere into a situation where it would have a brief hot summer and a long cold winter (with, however, the total heat received each half year being equal). Croll supposed that the critical factor was the situation in the northern hemisphere winter. If it were long and cold, the snow and ice would not all be melted during the following summer, even if the northern summer was receiving a higher amount of solar heat energy than the annual average.

The idea can be understood by reference to Figure 9.2. The various parameters represented in this figure—the shape of the ellipse, the obliquity of the ecliptic, and the positions of the equinoctial points—each vary over time with their own cyclical changes, giving a complicated overall pattern.

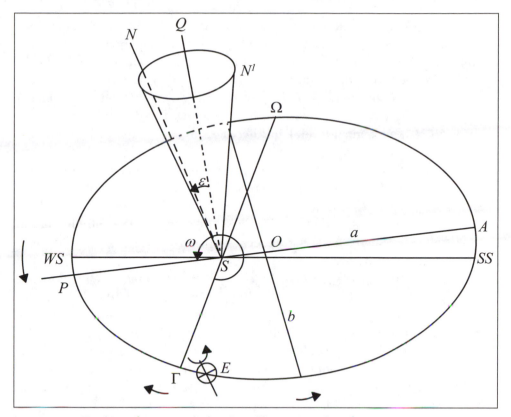

Figure 9.2: Earth's orbit around the Sun illustrating the phenomenon of precession. Equinox = one of the two days in a year when night and day are of equal length. A line can be imagined in space passing through the positions of the Earth (E) on the two days of equinox. This line points to two different positions on the celestial sphere, and the equinoctial line is defined thereby. It moves over the centuries in the so-called precession of the equinoxes. The circular "wobble" of the Earth's axis in space is responsible for the precession. (It is caused by the gravitational attraction of the Sun and Moon acting on the Earth's equatorial "bulge.") In the figure, S represents the Sun's position. O is the center of the Earth's elliptical orbit. a and b represent the major and minor axes of the ellipse, respectively. The ellipse defines the "plane of the ecliptic." P represents the Earth's position at "perihelion" (closest approach), and A is its position at "aphelion." The line SN is parallel to the Earth's axis of rotation. SQ is the perpendicular to the plane of the ecliptic through the center of the Earth's orbit. The angle NSQ is the same as the inclination of the Earth's spin axis to the plane of the ecliptic and is called the "obliquity of the ecliptic" (ε). WS represents the winter solstice (the shortest day of the year in the northern hemisphere). SS represents the summer solstice (the longest day in the northern hemisphere). Γ(gamma) and Ω (omega) represent the Earth's positions at the vernal (spring) and autumnal equinoxes, respectively, when lengths of night and day are equal. The line $\Gamma\Omega$ is the "equinoctial line." It points to particular points on the celestial sphere, but because of the slow wobble (precession) of the Earth's spin axis in space, $\Gamma\Omega$ slowly migrates around the ellipse. ω represents the angle of "longitude" of the direction of the perihelion relative to the vernal equinox. It moves as indicated in the figure. WS and SS correspond to the present positions of the solstices.

If the line ΓΩ were, at some time, directed along *PA*, then the heat distribution in the northern and southern hemispheres would be equal, whereas if it were at right angles to *PA*, there would be a short hot summer and a long cold winter in one hemisphere and vice versa for the other hemisphere. So there would (or could) be alternating glacials and interglacials in both hemispheres when the ellipticity is high, but if and when the orbit is nearly circular, as it is for most of geological time, the differences in the heat distribution effect would be negligible, and there would be no glaciations. So the theory seemingly provided just what was needed to satisfy the requirements of the stratigraphic column (episodic glacial epochs). Croll's image for his theory, as given in his *Climate and Time* (frontispiece), is reproduced in Figure 9.3. In his later work, Croll thought that glaciation was most likely when the obliquity of the ecliptic (ε) was small, for at such times the polar regions would receive less heat throughout the year.

Croll's critics made various objections. Some complained (like von Humboldt) that the quantity of heat falling on the Earth in either hemisphere was equal in winter and summer regardless of the orbit's shape or the orientation of the equinoctial points. Croll responded that (1) a large quantity of ice and snow formed during a cold winter would not melt during a short hot summer because of the "latent heat" needed to melt the snow and ice accumulated in the winters (so ice would grow from year to year), (2) large amounts of the Sun's radiant heat could be reflected back into space by a large ice cap, and (3) there could be substantial fog or cloud formation in a deteriorating climate that could prevent the Sun's energy reaching the Earth's surface. He also considered the effects of winds and ocean currents such as the Gulf Stream (Croll 1884)—which will not, however, be discussed here.

All this seemed satisfactory, and Croll's ideas were acclaimed during his lifetime. He was elected a Fellow of the Royal Society and awarded an honorary doctorate by St. Andrews University. When he retired early because of ill health, numerous eminent scientists lobbied successfully for him to receive a pension, contrary to the usual regulations.

Other authors offered somewhat different astronomical theories (e.g., Murphy 1869, 1876; Berger 1988), some stressing precession and others the obliquity of the ecliptic. The views of the Belfast manufacturer and amateur geologist Joseph Murphy (1827–1894) differed from Croll's theory in that he supposed that long cool summers and short mild winters favored glaciation (rather than long cold winters and short hot summers that Croll envisaged). That is, for Murphy, glaciation in a given hemisphere would occur when *winter*, not summer, occurred "*in perihelio*" and when the Earth's elliptical orbit was in a condition of high eccentricity.

Others doubted that the astronomical variations were sufficient to produce ice ages. However, in the United States, Gilbert (1895) (see p. 70) discussed certain regularly recurring Cretaceous beds in the Arkansas River basin and at the base of the Rocky Mountains—marine shales and sandstones with thin calcareous beds, with about 15 repetitions in all—which he thought could not

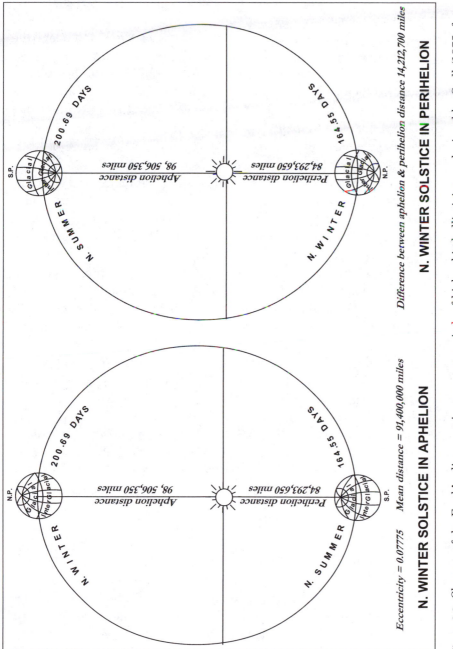

Figure 9.3: Changes of the Earth's alignment in space, at periods of high orbital ellipticity, as depicted by Croll (1875, frontispiece). The changes in alignment arise from the precession of the equinoxes and supposed concomitant differentiation of glaciation in the northern and southern hemispheres. Image redrawn by Ricochet Productions.

be accounted for by upheavals or subsidences, the shifting of waterways or divides, or changes of location of oceans. The phenomena appeared to be too regular. So he turned to the possibility that climatic changes due to astronomical causes might have been responsible. Various supporters of the astronomical theory were mentioned, including Croll. The changes might have been due to alterations in wind circulation, pole movements induced by the waxing and waning of polar ice caps (causing alterations in the Earth's center of gravity), or changing arid and humid conditions that altered vegetation and hence the acidity of waters, which could influence the type of sediment precipitated. Gilbert thought that his cycles coincided best with Croll's precession cycle.

Nevertheless, there was a fatal objection to Croll's theory that became fully apparent only after his death. As mentioned, the theory required that while one hemisphere was enduring glacial conditions, the other hemisphere enjoyed an interglacial. But as geological knowledge of the southern hemisphere increased, it became clear that this was not the case: the whole Earth suffered a glaciation at the same time. In addition, Croll had the last Ice Age finishing about 80,000 years ago, but geological evidence, such as the counting of varves (see p. 138), began to suggest that 10,000 years was more nearly correct. So Croll's theory was rejected for many years. People sought to account for glaciations in terms of changes in the energy emitted by the Sun itself; by the rise and fall of land at different latitudes, as Lyell had proposed; or in terms of different land bridges (like the Panama isthmus) that might or might not be present at different times, thus modifying the flow of ocean currents. (If the Gulf Stream did not flow today, northern Europe, especially Scandinavia, would be much colder than at present.) And after the advent of plate tectonics, ideas were floated about changes to ocean currents or regular winds, arising from the disposition of tectonic plates on the Earth's surface, that likewise might produce glacial conditions.

Meanwhile, evidence for more than one glaciation in the fairly recent geological past had been growing, through the work of such authors as the Scottish amateur geologist Robert Chambers (1802–1871) (1853); the professor of geology at Lausanne, Switzerland, Adolph Morlot (1820–1871) (1855); James Geikie; the geomorphologist Albrecht Penck (1858–1945) (see p. 75); and the climatologist and geographer Eduard Brückner (1862–1927).

One way to recognize multiple glaciations is to find different kinds of till (perhaps differing in color or texture, with different contained fragments, or perhaps separated by sands or gravels indicating deposition by flowing water during a period of milder climate). Geikie was able to find such evidences in Scotland and described them in his *Great Ice Age* (1874) and *Prehistoric Europe* (1881). One might also find evidence for glacials and interglacials by examining outwash gravels from mountain ranges, which might form distinct river terraces derived from moraine material.

River terrace studies were conducted by Penck in the 1880s, examining river valleys running into the Danube from the northeastern side of the Alps. Working for his doctoral thesis at Munich (*Die Vergletscherung der deutschen*

Alpen, 1882), he studied, for example, the Steyr Valley, where he observed what appeared to be four distinct sets of terraces, made of outwash gravels, one near the level of the present river and the other three at higher levels, evidencing separate glacial epochs. Each terrace could apparently be traced up the valley to the material of a terminal moraine. During a glacial epoch, there was strong erosion in the mountains, but not much debris was carried away because of the arid conditions. With the climate ameliorating, river flows increased and carried away previously deposited gravels, cutting deeper and wider beds (by river meandering). Then, after the next glaciation, there would be further deposition of gravel in the wide valley and so on. So terraces could give a record of successive glaciations.

In later collaboration with Brückner, Penck designated a series of four distinct terrace gravels and four corresponding Ice Ages that he named the Günz, Mindel, Riss, and Würm glaciations. The first three names referred to rivers running north from the Alps into the Danube, while the fourth was a tributary of the Isar River near Munich (Penck and Brückner 1901–1909, 1:110). Durations for the interglacials were suggested by analogy with the rates at which sediments appeared to be deposited in modern Swiss lakes, and this allowed an estimate of the time since the last glacial as 20,000 years and as 60,000 and 240,000 years for the Riss–Würm and the Mindel–Riss interglacials (Penck and Brückner 1901–1909, 3:1169). Corroboration and correlation appeared possible with North American glaciations also, where the so-called Nebraskan (oldest), Kansas, Illinoian, and Wisconsin (youngest) glaciations were proposed. Geikie toyed with five Ice Ages in Britain, but the German/Austrian ones carried the day, and Günz, Mindel, Riss, and Würm became a mantra for glaciologists or Pleistocene geologists in the interwar years. ("Pleistocene" is the name for the time span of the most recent series of Ice Ages, up to Recent times—i.e., the end of the last Ice Age or the beginning of the present Interglacial. The last glacial epoch for northern Europe ended about 10,000 years ago.)

But the Penck/Brückner theory fell apart after World War II. In 1948, the beginning of the Pleistocene was given by international agreement as 1.8 million years ago, corresponding with the appearance of cold-water species in southern Italy (see p. 163). Deep-sea rock cores revealed sediments showing evidence of many more glacials and interglacials than the classical four. From Bohemia, George Kukla, who later emigrated to the United States, reinvestigated Penck and Brückner's terraces and found that some of them contained interglacial materials. The terraces thus came to be regarded as being of tectonic origin rather than the products of climatic changes. The situation was much more complicated than the nineteenth-century authors or Penck and Brückner had imagined.

Nevertheless, theoretical work had been going on in the earlier decades of the twentieth century that in a sense took over from where Croll left off and allowed the post–World War II development of paleoclimatology, with a theoretical basis derived from astronomy. This new work was immensely more complicated and powerful than anything that Croll could have imagined, and it allowed a synthesis of theory and observation in a wonderful manner—all based on cycles! The work derived from ideas developed by the Serbian mathematician and astronomer Milutin Milanković (1879–1958) of Belgrade University.

Milanković took a Ph.D. at Vienna and returned to Belgrade, where he was appointed professor of applied mathematics in 1909 and spent his working life. As a young teacher, he pondered which field of study to choose, with a preference for something in applied mathematics that would have cosmic significance and would be directed toward a major unsolved scientific problem. He started with meteorology, which he found to consist mainly of endless empirical data without any mathematical backbone. But his reading led him to the problem of the Ice Ages. He approached it as had Croll but with the sophistication of a professional mathematician, seeking to calculate the periods of the different cycles in the Earth's orbit: the eccentricity of its ellipse (ellipticity), the tilt of its axis relative to the plane of the Earth's orbit around the Sun (obliquity of the ecliptic), and the precession of the equinoxes. From his calculations of these parameters, he wished to estimate the intensity of the total solar radiation striking the Earth (the total "insolation," or total incoming solar radiation). There was also the question of how the incoming radiation would be distributed over the Earth's surface through time.

As a Serb, Milanković was arrested by the Austro-Hungarian authorities during World War I and was then transferred to Budapest, where he was kept under police surveillance. But this gave him time to develop his ideas (rather like Einstein's period in a Swiss patent office). After the war, Milanković published the ideas he had developed during his period of arrest in *Théorie mathématique des phénomènes thermiques produits par la radiation solaire* (1920). It sought to estimate the Earth's surface temperatures due to solar radiation— which varied according to astronomical parameters—during the past 130,000 years and also the temperatures on the Moon and the other planets since there was interest in whether any of them might offer conditions where life might be possible.

Milanković's approach was similar to but different from Croll's. He sought to estimate the insolation at different latitudes and seasons and relate these to the total planetary heat balance, taking account of the reflection of solar radiation that might be caused by ice caps. He pointed out that the radiation arriving at the poles might be *greater* than that received at the tropics at the height of the polar summers, and if circumstances conspired to produce a cold summer, then *that* state of affairs—rather than a long winter—might cause the onset of glaciation. (In this regard, Milankovićs theory was more akin to Murphy's than Croll's.) So Milanković calculated the variation, over time, of the northern hemisphere insolation at three different latitudes and found various epochs when it dropped significantly.

Milanković's work attracted the attention of the distinguished Russian-born meteorologist and climatologist Wladimir Köppen (1846–1940) (director of the Hamburg Meteorological Observatory and later professor at Graz) and his (now) better-known son-in-law, Alfred Wegener (see p. 93), who were working on the problem of ancient climates. They invited Milanković's collaboration. Köppen believed that knowledge of the summer temperature of the Arctic was the key to understanding glaciations. He therefore asked Milanković to calculate the summer insolations at latitudes 55°N, 60°N, and 65°N for the past 650,000 years. By 1923, Milanković had worked out approximate solutions.

In 1924, Köppen and Wegener published *Die Klimate der geologischen Vorzeit* (*The Climates of Former Geological Times*), reproducing graphs for northern hemisphere insolation at three latitudes, as calculated by Milanković (see Figure 9.4). It is interesting that in his early efforts Milanković sought to recognize the Günz, Mindel, Riss, and Würm glaciations in his results (with some interglacials within these main glacial epochs). This work led on to Milanković's *Mathematische Klimatlehre und Astronomische Theorie der Klimaschwankungen* (*Mathematical Climatology and Astronomical Theory of Climatic Changes*, 1930).

Then, in 1938, Milanković published graphs for his calculations of summer-season insolation at 15°N, 45°N, and 75°N. The first two of these revealed the effect of the 22,000-year precession cycle, while the third revealed the dominance of the 41,000-year tilt cycle (obliquity of the ecliptic), which was capable of causing an Ice Age simultaneously in *both* the northern and the southern hemisphere. Thus, the problem that had dogged Croll's version of the astronomical theory of the Ice Ages was apparently overcome.

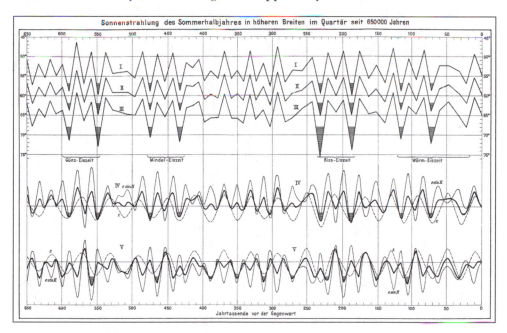

Figure 9.4: Milanković's radiation graphs for 55°N, 60°N, and 65°N, as published in Köppen and Wegener (1924, plate facing p. 256). ("Solar radiation for the summer seasons in the higher latitudes during the Quaternary for the past 650,000 years.") Note the attempted linkage to the four glacial epochs of Penck and Brückner (1909).

Milanković also attended to the possible movements of the poles, facilitated by his studying the theory of continental drift under Wegener and the distribution of the Earth's internal matter according to the principles of isostasy. Milanković then attempted to estimate the altitudes of the snow line at different latitudes in the recent geological past and the changes in the size and distribution of the polar ice caps according to changes in the snow lines. With an increase in their size, more sunlight would be reflected into space so that there would be a positive-feedback situation, leading to further cooling or amplification of the cooling effect that had been calculated from first principles by celestial mechanics.

Milanković continued to try to perfect his arguments, and his magnum opus was, he recorded, written in 539 days, being completed in 1941 as *Kanon der Erdestrahlung* (*Canon of Insolation*), published by the Royal Serbian Academy, thus bringing his life work and theory to a successful conclusion. A new translation has recently been published (Milanković 1998).

It is beyond the scope of this book to give an analysis of Milanković's mathematics, but some idea of the results of his thinking can be mentioned. There were, he showed, three main astronomical cycles involved in glaciation: the eccentricity of the elliptical orbit (a period of 98,000 years), the obliquity of the ecliptic (41,000 years), and the precession of the equinoxes (two effects: 19,000 and 23,000 years). These last arise from the "wobble" of the Earth's axis in space, accompanying movement of the Earth's elliptical orbit, *as a whole,* around the Sun.

As mentioned, with Köppen's support, Milanković defused the problem of the supposed alternation of glacial epochs in the northern and southern hemispheres by focusing attention on insolation at latitude 65°N (most of the Earth's land above sea level is exposed in the northern hemisphere) and assuming that cooling occurred *worldwide* if the northern hemisphere were significantly cooled. This could happen if there were appropriate feedback processes and if the northern hemisphere were cooled in response to astronomical causes. More snow would increase the reflection of sunlight, giving a cooler atmosphere. On the other hand, with cooling, water would be locked up in glaciers or polar ice sheets, causing the sea level to fall, thereby exposing the continental shelves. But this would allow increased vegetation, tending to absorb rather than reflect radiation, whereas fogs produced by cooling would prevent solar radiation from reaching the Earth's surface. The production of sea ice would, on the other hand, increase the reflection of radiant heat. There could also be an increase in the fresh surface water in the oceans, which might "turn off" the so-called thermohaline circulation of the oceans' currents and thus "switch off" the Gulf Stream—which would produce catastrophic cooling in Europe were it to happen today. (Today, there is a gigantic current, linking the Atlantic, Indian, and Pacific oceans, with both shallow- and deep-water components, involving the movement of warm shallow water and denser, more saline, cool water. It is sometimes called the global conveyor belt.) However, in a cold climate there would be less photosynthetic activity and hence more carbon dioxide in the atmosphere (greenhouse effect) and less evaporation from the colder and smaller seas. Hence there would be less precipitation of snow and glacier formation. As yet another factor, the weight of the ice sheets would tend to depress land surfaces by isostasy, and this would diminish cooling. There were many relevant factors, not all acting in the same direction.

Alternatively, if the astronomical forces were tending to increase the solar radiation, this would melt some ice and snow and reduce the reflection of solar radiation from the Earth's surface. The thermohaline current might be restored, leading to climatic amelioration in northern Europe by courtesy of the Gulf Stream. Sea levels would rise.

As mentioned, as long as the greater part of the Earth's land surface was in the northern hemisphere, the global climate could be dominated by the situation in the northern hemisphere. Of course, if (by continental drift) it came about that a greater proportion of the land was in the southern hemisphere, then *that* would control the Earth's glacials or interglacials or whether there was an *Eiszeit* at all. It should be noted that while Croll's theory gave greatest weight to changes in the ellipticity of the Earth's orbit, Milanković thought the tilt of the Earth's axis ("obliquity of the ecliptic") was the dominant factor.

Despite its virtues, Milanković's theory was largely ignored or regarded as falsified for some fifty years after he launched it in the 1920s. This may have been due to his association with Wegener (whose ideas were unpopular), the sheer difficulty of Milanković's mathematics (for most geologists) and the great complexity of his problem, or the fact that it was put forward by someone in a remote part of Europe and chiefly in German or Serbian. But there were also empirical objections. There was, for example, a discrepancy between the strong global cooling predicted by Milanković's theory at 25,000 years before the present (BP) (from estimated northern-hemisphere insolation at 65°N) and an actual cooling event that apparently occurred at 18,000 years BP. Moreover, the correlation between Milanković's four main cold epochs and the four Ice Ages of Penck and Brückner was worth little when the latter theory was discredited by Kukla. Moreover, the postwar invention of carbon-14 (^{14}C) dating methods by Willard Libby (1908–1980) (1952) and his students at the University of Chicago and their application to the dating of organic matter in fairly recent glacial and interglacial deposits in the United States suggested that a warm period occurred 25,000 years ago, whereas Milanković's calculations suggested a glacial epoch at that time. Other results seemed equally unsatisfactory. But that did not, in the event, prove to be the end of his theory.

In 1955, the Italian investigator Cesare Emiliani (1922–1995) reported a correlation between Milanković curves and ^{18}O and ^{16}O ratios in plankton sampled in deep-sea cores of carbonate oozes at 10-centimeter intervals. (Water containing the ^{18}O isotope is denser than water containing ^{16}O. Therefore, there is a higher proportion of $H_2{}^{18}O$ in the seas in cold conditions than in warmer conditions, and more is taken up by the plankton.) Moreover, he found evidence for *more* than the four classic Penck glaciations: seven in the Caribbean cores and 15 in the Pacific cores. Emiliani moved from Italy to Florida in 1957 and was instrumental in getting a program of seafloor coring under way, known as the Joint Oceanographic Institutions for Deep Earth Sampling (JOIDES), which collected evidence that bore on both the history of climate changes and plate tectonic theory.

Then, in 1968, Wallace Broecker of the Lamont Geological Laboratory and his coworkers (Broecker et al. 1968) reported their work on the study of the reef terraces around the island of Barbados. These were older than the time range of the ^{14}C method but could be investigated by studies of the radioisotopes of uranium, thorium, and potassium, and it appeared that the sea stood at high levels 80,000 and 120,000 years ago, as predicted by Milanković—but also at 105,000 years, which was not predicted by him. Broecker found, however, that if one examined insolation calculations for lower latitudes (45°N) rather than 65°N, as calculated by Milanković, the precession cycle was more prominent and the Barbados data could be accommodated. The results were later confirmed for New Guinea terraces.

In 1976, a landmark paper was published in *Science* by James Hays, John Imbrie, and Nick Shackleton. It analyzed cores from the southern Indian Ocean, which yielded a climatic record extending back some 450,000 years, with sediment estimated to have accumulated at about 3 centimeters per 1,000 years. Besides estimating the oxygen isotopic composition in plankton as Emiliani had done, they also estimated summer surface-water temperatures over time by examining the remains of radiolarians (with characteristic silica skeletons) and particularly the relative abundance of *Cycladophora davisiana*, which is characteristic of cold-water conditions. The investigation revealed a close relationship between the cycles predicted by Milanković theory and the changes observed in the submarine stratigraphic record. (The analysis involved "unraveling" the signals from different cyclic processes, which appear as a "composite" curve in the isotope records, a procedure called "spectrum analysis".) There was a time lag between the two series of curves, but the three main Milanković cycles were recognizable from the core data: the (approximately) 100,000-, 42,000-, and 23,000-year cycles, though the correlations were not precisely one to one.

From the publication of that paper on, the "reality" of Milanković cycles has been largely accepted. *Nature* published an article on the topic by André Berger of Louvain Catholic University (1977), albeit with some modification of Milanković's work for the ellipticity cycle of about 100,000 years. Berger's plots for the three main components of orbital eccentricity, precession, and obliquity of the ecliptic are shown in Figure 9.5, and work continues to the present to produce ever-improved fine-tuning of the data (Hinnov 2000). Conferences are now held specifically devoted to "Milanković work" and the "articulation" of what has become a (or *the*) "Milanković paradigm" (see Serbian Academy of Sciences and Arts 2004). We may say that the problem of the Ice Ages is now solved in its essentials (Imbrie and Imbrie 1979), the hard theoretical work having been accomplished by Milanković, with geologists gradually finding the necessary supporting empirical evidence. But whether Milanković theory provides an explanation of all the issues raised in the following two chapters is another matter. As one recedes further into the past, the backward extrapolations of modern astronomical data become more and more insecure. One could begin to argue in a circle (!) if the stratigraphic information available in

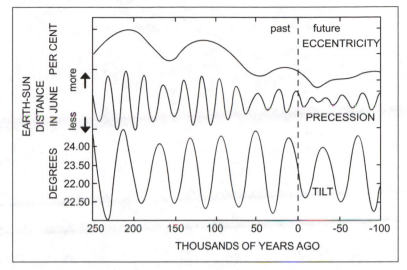

Figure 9.5: Components of the variations of the Earth's motion in space, through time (Berger 1988, 636). © 1988. Reproduced by permission of American Geophysical Union. Curves developed from calculated numerical data published in Berger (1978). Image redrawn by Ricochet Productions.

the rocks for, say, the Mesozoic or the Upper Paleozoic were used to try to aid the calculation of the conditions of the Earth's orbit in those far-off times or vice versa.

10

CYCLES IN THE STRATA: CYCLOTHEMS

It is a commonplace observation that sediments form layers, sometimes with quite abrupt changes in the type of rocks involved, such as shale being over-lain by sandstone. And it is no less common to find apparent breaks in the deposition of sedimentary rock as "bedding planes." For sediment accumulation, there are necessarily a source and a basin into which sediment can be deposited. The formation of substantial thicknesses of water-transported sediment demands either the sinking of the basin, a rise of sea (or lake) level, or some combination of the two. With various possible causes for such changes, a thorough explanation of the features of sedimentation and sedimentary deposits can be a complicated matter. But many features of the sedimentary record suggest cyclic processes.

In attempting to defend his evolution theory, Darwin (1859) sought to account for the abrupt changes in fossil forms found in the stratigraphic record by emphasizing the incompleteness of the geological evidence. He recalled that the coast of Chile, visited during his *Beagle* voyage, contained few Tertiary deposits despite the ample supply of sediment from the rising mountain ranges of the Andes close by to the east. The reason, he explained, was that any Tertiary deposits that had been deposited were easily removed by erosion of the rising land surface. He continued,

> All geological facts tell us plainly that each area has undergone numerous slow oscillations of level, and apparently these oscillations have affected wide spaces. Consequently formations rich in fossils and sufficiently thick and extensive to resist subsequent degradation, may have been formed over wide spaces during periods of subsidence, but only where the supply of sediment was sufficient to keep the sea shallow and to embed and preserve the remains before they had time to decay. On the other hand, as long as the bed of the sea remained stationary, thick deposits could not have been accumulated in the shallow parts,

which are the most favourable to life. Still less could this have happened during the alternate periods of elevation; or, to speak more accurately, the beds which were then accumulated will have been destroyed by being upraised and brought within the limits of coast-action. (Darwin 1859, 291–92)

So, with "oscillations," there would be frequent breaks in the stratigraphic record.

We need not attend here to other nineteenth-century ideas but may turn immediately to the consideration of rhythmic sedimentation by the Yale geology professor Joseph Barrell (1869–1919), who in an influential paper examined the reasons why the sedimentary record is commonly very incomplete (Barrell 1917).

Like many other American geologists, Barrell was influenced by Davis, so that he saw erosion as "essentially a pulsatory process," with one of Davis's cycles of erosion being a "single rhythm." The notion of "base level" was essential to Barrell. Sediment deposition, he said, occurred just below a local base level and could proceed by upward movement of the base level or downward movement of the floor of the sediment basin. Erosion cuts down to base level, while deposition builds up to it. So sedimentation was determined by changes in base levels. Barrell accepted that sediment loading would depress a basin, but for sedimentation to begin, initial subsidence was required. Sediment accumulation in a basin is discontinuous, as sediments are deposited and scoured away by currents and the hollows filled by the arrival of further sediment. The minor breaks in deposition Barrell called "diastems." He accepted the idea that sedimentation rhythms might be correlated with "Crollian" climatic cycles.

Barrell went on to consider denudation processes and pointed out that they occurred more actively the greater the angle of slope of the surface undergoing erosion. Although governed by changes in base level, deposition necessarily occurred on a sloping surface also, and with a subsiding basin, deposition must occur on a gradually tilting surface.

Barrell then proceeded to argue that the stratigraphic record must *necessarily* be incomplete, even when large gaps were not evident in the form of unconformities. He imagined sedimentation in a generally sinking basin, but deposition might be disturbed by oscillations in the subsidence or minor rises and falls of sea level. There could be many contributory causes to the fluctuations in the general subsidence: sediment loading, sea-level lowering due to climatic changes, isostatic rebound following the melting of an ice cap, the operation of general tectonic forces, changes of base level due to erosion of land surfaces and/or deposition of sediments, and so on. The operation and effect of such factors was summarized diagrammatically (Figure 10.1).

In this figure, the line A–A represents a general rise in base level accompanying the deposition of sediment in a subsiding basin. The rate of subsidence is initially slow, increases to a steady rate, and then tails off again at the right-hand side of the figure. The "wavy" line B–B represents hypothetical minor

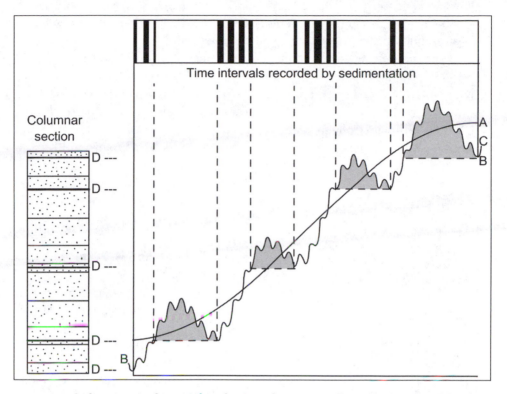

Figure 10.1: Sediment record in a sinking basin undergoing oscillations in base level (Barrell 1917, 796). The added shading indicates the quantities of sediment that were initially deposited and then subsequently removed. Image redrawn by Ricochet Productions.

oscillations superimposed on the general subsidence. The left-hand part of the figure represents the ("surviving") sediments that accumulate in the basin. Minor breaks in the sedimentation are represented by thin horizontal lines across the hypothetical stack of accumulated sediment. More substantial breaks (diastems [D]) are represented by thicker lines. The top part of the figure represents the times (thick black vertical lines) that are physically represented in the surviving sediment that is figured in the vertical sedimentary sequence. Note that all the sediment deposited in the shaded areas is subsequently removed by erosion, as there is a cyclic variation in the rise of the base level (which, however, has an overall upward trend). The same holds true for each individual minor change in the base level (the small "waves" superimposed on the larger "waves," which are themselves superimposed on the large-scale S curve.

It will be seen, then, that only a fraction of the sediment initially deposited remains in the record for geologists' inspection—and much more may be lost in subsequent orogenies and larger-scale erosion of land surfaces. Thus, Barrell's model supported Darwin's contention that the sedimentary record is substantially incomplete. When we look at a cliff face of sediments, only part of the total time from the beginning to the end of the deposition process is represented. We note that Barrell's argument depended on the cyclic processes of erosion and deposition.

With such considerations in mind, let us turn to thoughts about sedimentation as subsequently developed in the United States in relation to the mid-continent areas containing repetitive sequences of almost horizontal rocks, associated with the coal measures of the Mississippian and Pennsylvanian (which approximate to Europe's Lower and Upper Carboniferous). The American strata extend from the coalfields of Pennsylvania and Virginia in the east; through Ohio and Illinois; through Iowa, Nebraska, and Missouri; and down through Kansas and Oklahoma into Texas. In many areas, such as Illinois and Kansas, the beds are almost horizontal and of great extent, with perfect "layer-cake" bedding of sandstones, shales, coals, and limestones. Somewhat analogous horizontally bedded strata can be found elsewhere, as in the Pennines of northern England.

These regular sequences came to be called cyclothems (Greek *thema* = something laid down), and their significance and theoretical interest were recognized as the detailed mapping of the central United States was pursued during the first half of the twentieth century. The first person to write about them specifically was Johan Udden (1859–1932), who migrated to the United States from Sweden when young and studied at the University of Minnesota and Augustana College, Illinois, where he subsequently taught. He also worked for some of the state surveys and eventually moved to the University of Texas's Bureau of Economic Geology, where he became director in 1915. While working in Illinois (1906–1911), Udden mapped the "Peoria Quadrangle" and was the first to describe the distinctive repetitive character of Illinois' Pennsylvanian sediments (Udden 1912) (Figure 10.2).

Udden (1912, 49) took the view that the cycles represented "recurrent submergences, alternating with periods during which the sunken areas were filled to the level of the surface of the sea," and that "the four cycles [that he recognized] represent[ed] recurrent

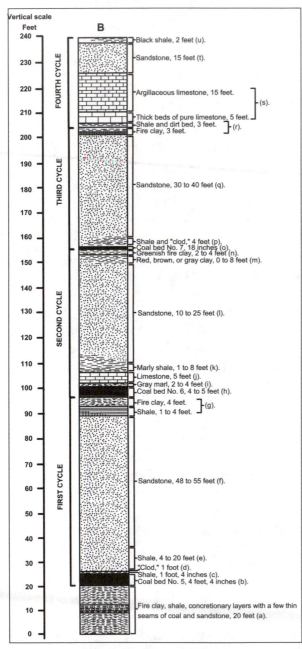

Figure 10.2: Sedimentation cycles (Udden 1912, 27).

interruptions in a progressive submergence." He was struck by the horizontality of the beds, their uniform texture, and their wide extent. He supposed that each succession began with a coal swamp, which became inundated by the sea with the formation of limestone. Then sand was swept in, filling the sea to the surface and forming sandstone, the subaerial weathering of which produced clay, on which the next coal swamp formed and so on. His suggestions were not pursued by others at that time.

Later, however, they were taken up with energy by the Kansas Survey's director, Raymond Moore (1892–1974); by Harold Wanless (1898–1970) of the University of Illinois; and by Marvin Weller (1899–1976) of the Illinois Survey (subsequently the University of Chicago).

As a young geologist who had studied under Chamberlin and Rollin Salisbury at Chicago, Moore started work on the correlation of Permian and Pennsylvanian rocks in the mid-continent, working his way south from Nebraska to Oklahoma. In his paper "Environment of Pennsylvanian Life in North America," Moore (1929) described the formations' various life forms, both plant and animal. He also suggested what their paleoenvironments might have been. He noted the "interfingering" of marine and terrestrial or nonmarine facies and suggested that the topography had been such that small changes in the levels of land and sea could have produced large changes in the positions of the shorelines. These could have been produced by repeated downwarpings of the sediments basin(s) and variation in sediment supply.

Suess (1904–1924, 4:80) had suggested—not on the basis of firsthand experience, of course—that the mid-American sediments had been deposited on a surface with their present gentle westward dip. But Moore rejected this, as it required numerous quite substantial elevations and subsidences to produce the observed widespread distribution of similar strata. Instead, he preferred the idea of slow subsidence of a shallow sea that gradually overlapped to the west of the basin. There was, he suggested, a "vast swampy plain slowly built up and extended by stream alluviation, periodically more or less widely submerged to shallow depth in the sea" (Moore 1929, 487). The stratigraphic record would probably have been relatively complete (cf. Barrell's mode of analysis), as presumably relatively little material would have been lost to erosion, though sediment could always be shifted by scouring.

Writing the following year, Weller (1930) reviewed the idea of sea-level changes as being responsible for the phenomena and regarded all of them as the product of cyclical changes. (He preferred the term "cycle," implying recurrence, to "rhythm," and this usage seems to have prevailed.) He suggested that a typical cycle produced the following:

Marine

8. Shale, containing "ironstone" bands in upper part and thin limestone layers in lower part
7. Limestone

6. Calcareous shale

5. Black "fissile" shale

Continental

4. Coal

3. Underclay, not uncommonly containing concretionary or freshwater limestone

2. Sandy and micaceous shale

1. Sandstone

Unconformity

Using this "paradigm" sequence of sediments for a cycle, Weller sought to make cross correlations over a wide geographical range of deposits. Where the deposits were not physically connected, this was ascribed to local downwarping that had preserved patches of original sheets of sediment, whereas other parts had been lost to erosion, subsequent to deposition. The "idiosyncratic" features of any particular sedimentary cycle should enable it to be recognized wherever it appeared. Like Udden, Weller thought the black shales represented *shallow*-water deposits.

Weller also attended to the cause(s) of it all. There had, he supposed, been repeated periods of widespread subsidence, followed by uplift, which brought a cycle to a close. Thus, he was an advocate of diastrophism as the prime cause: climate change could be omitted from the explanation.

In 1931, the Illinois Survey held a Quarter Centennial Celebration symposium in which issues connected with cyclic sedimentation in the Pennsylvanian were canvassed. Concentrating on Kansas, Moore (1931) reviewed the issues and emphasized that there was a continuity of strata across the great Pennsylvanian sediment basin so that correlation was possible as a result of the convenience of the cyclic sedimentation being repeated at different localities with "remarkable fidelity." He remarked that the thickness of the sediments increased to the south and that the source of the sediments appeared to be from the south. He inclined to his earlier explanation (1929) of the origin of the sediments, regarding the black shales as shallow-water deposits.

Weller (1931) reiterated his interest in earth movements as the cause of the cyclicity and described what he called the "diastrophic cycle," which could have proceeded as follows. There was initial regional uplift of North America, yielding a newly formed land surface, which was then dissected by the forces of erosion, with entrenching of river valleys. With continued uplift of Appalachia, these valleys became filled with sediment that spread out on a vast alluvial plain in mid-America. The uplift of Appalachia was compensated by a lesser subsidence in the interior, where shallow basins were formed in which freshwater limestones accumulated and sediments from higher ground were also received. Drainage was poor. Further slight subsidence led to luxuriant plant growth and swamps in which peat accumulated, later converted to coal. Further subsidence allowed the advance of the sea and the accumulation of black muds with algae, which later became black shales. On further subsidence, the seas cleared and calcareous sediments formed, which subsequently became limestones. Then further uplift of Appalachia renewed sedimentation in the basin. Additional and stronger uplift of Appalachia caused withdrawal of the sea from the basin region(s), and thence a new cycle began. The whole theory relied on earth movements rather than climate changes. The long hand of Davis was evident in Weller's thinking.

Concentrating on western Illinois, Wanless (1931) emphasized the importance of correlations and drew several diagrams showing the correlation of sediment logs (e.g., Figure 10.3). Fifteen sedimentary cycles were supposedly evident in the basin. Causes were not really discussed, but for an area with important coal deposits, issues of correlation were more "interesting" than theoretical explanations of the sedimentation.

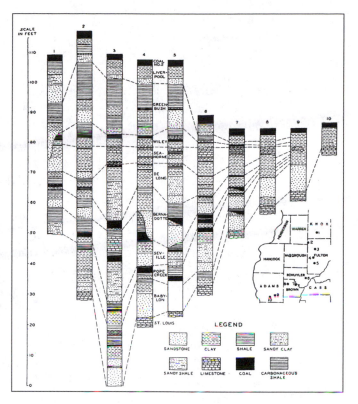

Figure 10.3: Correlations of strata in western Illinois (Wanless 1931, 189).

The following year, Wanless and Weller (1932) combined to propose a definition or description of a "typical" cyclothem, at the same time formally introducing the term:

> The word cyclothem is . . . proposed to designate *a series of beds deposited during a single sedimentary cycle* of the type that prevailed during the Pennsylvanian Period. A cyclothem ranks as a formation in the scale of stratigraphic nomenclature. (Wanless and Weller 1932, 1003; emphasis added)

The "typical" or "ideal" cyclothem was as follows. (But in many exposures, only parts of the whole could be seen, and all components were not necessarily deposited.)

Marine

Sand (top)
Thick shale
Thick limestone
Carbonaceous shale

Freshwater

Coal
Underclay

Freshwater limestone
Sandy shale
Fine-grained sandstone

The "pattern" reflected Wanless's (1931) thinking, discussed previously. The repetitive deposition of sediments in cyclothems had, of course, been known to coal miners for centuries, and the phenomenon had been remarked on by geologists in northern Britain back in the 1830s.

Wanless and Weller continued to explain the cyclic phenomena in terms of earth movements. But changes in sea level might also be invoked, as was done four years later by Wanless and Shepard (1936). Francis Shepard (1897–1985) had been a student of Chamberlin and other notables at Chicago, where he developed a reputation for being somewhat iconoclastic with regard to Chamberlin's ideas on diastrophism. Shepard thought that sea-level changes should be invoked to account for various sedimentary phenomena, rather than earth movements being the dominant factor. In his later career, he moved to the Scripps Institution of Oceanography in California, where he conducted pioneering surveys of submarine canyons, referred to in Chapter 8. Thus, he became a marine geologist and has even been called the "father of marine geology."

It is unsurprising, therefore, that Wanless and Shepard began to emphasize sea-level changes in cyclic sedimentation. In 1934, Wanless had extended his studies of cyclic sedimentation in the Pennsylvanian to the Rocky Mountains, further west than his previous studies, and this had given him a new perspective. Questions arose as to how there could be repeated upward and downward movements of different parts of mid-America with each subsidence exceeding the previous uplift and each rise and fall being accompanied by rises and falls of the adjoining areas supposedly generating the sediments. It was odd, also, that similar movements might have occurred at about the same time in the Carboniferous strata of Europe. And with all the supposed diastrophic movements, it was strange that so many of the sedimentation cycles in mid-America produced strata that remained almost horizontal in many places through to the present.

So changes in sea levels now seemed a preferable proposition to Wanless and Shepard. They could be caused by large-scale earth movements, almost anywhere on the globe, that led to changes in volume of the ocean basins; by changes in precipitation or the quantity of water contained in the atmosphere (which is temperature dependent); or the locking up of water in ice caps and glaciers. Glaciation was by then recognized in the Permian in several parts of the world. Wanless and Shepard were inclined to suppose that there might have been a series of glacials and interglacials in the Permian and Carboniferous, somewhat analogous to those in the Pleistocene, and this might account for some of the North American cyclicity. There was, however, a problem in that cycles were also found in the so-called Piedmont Terrain of New England, where they were thought to offer nonmarine facies and could not be accounted for by changes in sea level alone. These strata, however, contained deltaic sediments, and changes in sediment type might have had climatic causes, without the direct agency of the sea.

Having reviewed the possibilities, Wanless and Shepard opted for a glacial-eustasy explanation of the sedimentary cycles of mid-America (and elsewhere) arising from climate changes driven by astronomical forces or variations in the Sun's radiant energy. (Eustasy can be defined as global changes of sea level caused either by volume changes of the ocean basins or by changes in the volume of the oceans—from the locking up or release of water by the ice caps due to climate changes.) Moore had envisaged "megacyclothems" (cycles of cyclothems), which might, perhaps, correspond to the largest of the possible astronomical cycles (changes in the ellipticity of the Earth's orbit). The role of diastrophism was not excluded, but it was not regarded as the dominant factor. As there was no independent evidence of glaciation to account for the supposed eustatic sea-level changes during the Pennsylvanian (when cyclothem formation was most pronounced), the Wanless and Shepard theory was regarded as somewhat "hopeful." It should be noted, however, that *if* astronomically induced eustasy were the main cause, *then* it would allow the establishment of *global* stratigraphic correlation—such as Suess had contemplated (see p. 89). But the cause would be different from what he had suggested.

Moore (1936; Moore et al. 1944) began to put together data on cyclothems across the United States. He emphasized the lateral extent of individual cyclothems in Kansas and suggested that there had been major fluctuations in shoreline on which minor advances and retreats were superimposed. In central states such as Illinois, there appeared to have been alternations of marine (shales, sandstones, and limestones) and continental sedimentation (coal and so on). To the east, in the Appalachians, the sediments were primarily continental though still cyclic. And to the west, in Arizona and New Mexico, sedimentation was chiefly marine, though again yielding cycles of different types of sediment. During this period and in the following couple of decades, there was much discussion of cyclothems and megacyclothems and attempted correlations across different states; but the details need not detain us here. For further information, see Heckel (1984) and Dott (1992).

In 1942, Bowen Willman (1901–1984) and Norman Payne again sought to specify what an ideal cyclothem looked like and offered a diagram (see Figure 10.4). The components were not quite the same as those of Wanless and Weller (1932). But the principle of there having been cyclic sedimentation was not an issue. On the other hand, while accepting cyclicity, various types of cyclothem were proposed (Weller 1942, 1958). Such "articulation of a paradigm" (Kuhn 1962) might be expected in geology. A model or "exemplar" had to be "adjusted" in the face of more refined empirical evidence. But for present purposes, the issue was the general *explanation* of the phenomena of cyclic sedimentation, not the details in different localities.

A problem with the Wanless and Shepard (1936) idea was that it proved difficult or impossible to correlate the glacial epochs that they thought might be

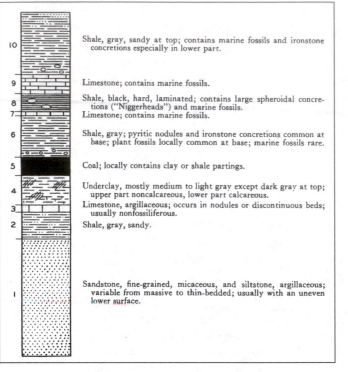

Shale, gray, sandy at top; contains marine fossils and ironstone concretions especially in lower part.

Limestone; contains marine fossils.

Shale, black, hard, laminated; contains large spheroidal concretions ("Niggerheads") and marine fossils.
Limestone; contains marine fossils.

Shale, gray; pyritic nodules and ironstone concretions common at base; plant fossils locally common at base; marine fossils rare.

Coal; locally contains clay or shale partings.

Underclay, mostly medium to light gray except dark gray at top; upper part noncalcareous, lower part calcareous.

Limestone, argillaceous; occurs in nodules or discontinuous beds; usually nonfossiliferous.

Shale, gray, sandy.

Sandstone, fine-grained, micaceous, and siltstone, argillaceous; variable from massive to thin-bedded; usually with an uneven lower surface.

Figure 10.4: Ideal cyclothem (Willman and Payne 1942, 86). Reproduced by permission of Illinois State Geological Survey.

responsible for global falls in sea level with other evidence of glaciation. So their idea was criticized by Weller (1956), who still favored diastrophic control of cyclicity. In response, Wanless went off to look at the evidences for southern hemisphere glaciation and maintained that the southern Paleozoic glaciations could have been responsible for the northern hemisphere cyclic sedimentation and the formation of cyclothems (Wanless and Cannon 1966).

But the north–south correlation seemed far-fetched to some commentators, and other explanations had been or were sought. Kingsley Dunham (1950) (subsequently director of the Geological Survey of Great Britain) had attributed the Yorkshire cyclothems to delayed isostatic adjustment to accumulating loads. Derek Moore of the Bureau of Mineral Resources, Canberra (who was familiar with cyclic deposits in Yorkshire, United Kingdom, and had written on them for his Ph.D.), suggested that the cause of cyclothem formation might be "delta switching" (Moore 1958, 1959) (cf. Figure 8.12). In the Carboniferous rocks of Yorkshire, he recognized eight major cyclothems, each with a basal limestone. He imagined the existence of a shallow sea in which limestone formed and a large delta (analogous to the Mississippi's) that deposited sediment in the sea. The marine basin was supposedly sinking, and the delta plain might fall below sea level, leading to the abandonment of the old delta and the initiation of a new one. The old delta would be inundated and receive a deposition of limestone and so on. Moore's ideas, though not satisfactory for explaining widespread sedimentary correlations, did attract some interest in the United States, such as John Ferm's (1975) work in the Appalachians. Ferm's paper is interesting in that it offered a diagrammatic section of the typical sedimentary deposits of the North Appalachian Plateau that is reminiscent of the figures that came to characterize the work of sequence stratigraphers (see Figure 11.7). Ferm's section looked like a flying duck seen in profile, so he dubbed it the "Allegheny Duck." Unfortunately, the name did not, as far as I'm aware, catch on!

Another suggestion was that of Thomas Robertson (1952) of the British Survey. Robertson proposed that plants would build up in a subsiding swamp until an area became flooded by the breaking of a sandbar. But such a local event could hardly account for the widespread, uniform, and almost horizontal cyclothems of mid-America. Alan Wells (1960) of Shell International supposed that worldwide sedimentation would displace sea water upward, but this would be countered by the down-warping of seafloors due to sediment loading. With the two processes taking place in tandem, there could be an oscillation of the absolute sea level, and a gradually subsiding basin could produce cyclic sedimentation.

It should be remarked here that correlations of strata at a distance, supposedly evidencing a general cyclicity of sedimentation in a region, can be in error because of observers' subjectivities, as was demonstrated in an often-quoted psychological experiment by Edward Zeller (1964). A group of graduate students was given plots of three different well logs, with various lithologies represented in the vertical sections, and they were asked to match these. After about five minutes, all had done so to their satisfaction. The students were then given a log of a mine shaft near Kansas City and were asked to match this with the three logs previously correlated. Within a few minutes, this task had also been accomplished. Sequences were cross correlated by matching the figures, assuming that the sediments exhibited cyclicity. *But*, Zeller had prepared his first three well logs on the basis of numbers drawn from the Kansas City telephone book! This was a cruel trick, but it revealed the possibility of subjectivity in matching stratigraphic data. Stratigraphers have subsequently taken precautions, using statistical analyses, to ensure that their matches reveal correlations that are based on more than casual similarities. Things may not be as they appear to the prepared mind!

Noting that cautionary tale, we return to consider the subsequent efforts to achieve explanations of the production of cyclothems in terms of sea-level changes. Such work has been spearheaded by Philip Heckel (b. 1938) of the University of Iowa, who in a series of papers (still ongoing) has sought to show that the Upper Paleozoic cyclothems of mid-America may be correlated with changes driven by astronomical causes, or Milanković cycles (Heckel 1984, 1986, 1994, 1995). One of his main innovations (Heckel 1977) was the suggestion that the dark fissile carbonaceous shales in the cyclothems should not be regarded as the product of swampy or shallow-water environments. Rather, he suggested, they were produced in relatively *deep water* (perhaps 200 meters) in anaerobic conditions. He envisaged the upwelling of cool, low-oxygen, nutrient-rich water, which produced algal blooms at the surface that could then sink to a phosphorus-rich seafloor, providing materials for the formation of carbonaceous black shales. Thus, black shales represented marine transgressions rather than regressions. This altered the whole picture as previously understood.

In a paper initially delivered at Princeton University in 1985, Heckel (1986) sought to estimate the time taken for the formation of a Kansas City cyclothem. He represented his interpretations graphically, but his model is here depicted in simplified form in the following table:

| "Outside" [nearshore] Shale | Shale, blocky mudstone | Offshore pro-delta; soil | (Sea level low) |

Upper Limestone	Oolite, calcareous sandstone/mudstone	Formed in shoals	(Regression)
Core Shale	Grey shale over black shale	Formed offshore	(Sea level high)
Middle Limestone	Calcareous mudstone	Formed offshore	(Transgression)
"Outside" [nearshore] Shale etc.	Shale, mudstone, coal (Regression)	Formed nearshore, terrestrial	(Sea level low)

The sort of rocks he was examining can be seen to advantage in various road cuttings and quarries near Kansas City (see Figures 10.5a and 10.5b, identification of units by Tom Stanley).

Heckel, then, examined a "pile" of 20 "major" cyclothems plus 14 incomplete intermediate cycles and 21 incomplete minor cycles and claimed that the whole could be correlated with European strata, extending over a known time range (about 8 to 12 million years). Moreover, he suggested, the various

Figure 10.5a: Cyclothems (Holliday Drive, Kansas City, by the southern bank of the Kansas River, near its junction with the Missouri River). Lane Shale (grassed top); Wyandotte Limestone; *Liberty Memorial deltaic shale; Iola Limestone (with included dark "core shale" [Muncie Creek Shale Member])*; Chanute deltaic shale (bottom). The limestones represent marine environments during either transgressions or regressions; the thick gray shales represent deltaic conditions. The dark ("core") shales within the limestones represent the circumstances of maximum water depth. The lithological names refer to different formations of the upper part of the Kansas City Group, in the Missourian Stage, of the Upper Pennsylvanian Series, of the Carboniferous Period. The so-called Iola Cyclothem is indicated in italics. It is overlain by the Wyandotte Cyclothem. Author's photograph, 2004.

major, intermediate, and minor cycles might be correlated with Milanković cycles. Thus:

Cycle Types	Estimated Duration of Cycle	Milanković Period	Astronomical Cycle
Major	235,000–393,000 years	413,000 years	Eccentricity cycle
Intermediate	118,000–197,000	95,000–136,000 (average 100,000)	Eccentricity cycle
Minor	44,000–118,000	41,000	Obliquity cycle
	None	19,000 and 23,000	Precession cycles

The figures were obviously imprecise, and the ranges in the second column resulted from assuming 8, 10, or 12 million years for the deposition of the total amount of sediment. Nevertheless, it seemed there was an empirical basis for believing that the mid-American cyclothems could have been produced by astronomical causes ("orbital forcing"). The changes in the Earth's orbit had affected the climate, which had caused glaciations (somewhere), which had caused sea-level changes, which had caused the formation of sediments as observed in mid-America! Heckel added a note from a referee at the end of his paper: "A most significant paper that makes giant strides between the stratal record and major Earth events." It seemed that geology and astronomy were becoming integrated most satisfactorily.

Figure 10.5b: Limestone with shale partings (Holliday Drive, Kansas City). The shale partings represent minor regressions, equivalent to parts of what Heckel referred to as intermediate or minor cycles. Author's photograph, 2004.

However, as might be expected (!), Heckel encountered criticism. George Klein and Debra Willard (1989) of the University of Illinois at Urbana–Champaign thought that tectonic flexure as well as sea-level changes could be involved. Klein (1990) further objected that the calculation of Milanković cycles so far into the past was fraught with difficulty and also that other time scales had been developed for the Pennsylvanian that were shorter than the one deployed by Heckel. All this would nullify his claims. There were subsequent exchanges on the issue that will not be pursued here. Heckel seems to have had the last word.

So a link was forged between astronomical cycles and ones long known to students of stratigraphy, sedimentology, and paleoecology. Nevertheless, caution is called for. As argued by Miall and Miall (2004a), there is a risk of circular argument. When one has empirical data from which are extracted mathematically the frequencies of the constituent cyclic processes that give rise to the data (spectrum analysis), there can be a tendency to ascribe discrepancies between what is observed in the sedimentary record and what is calculated à la Milanković to various ad hoc factors. Pattern matching can be subjective, as previously noted.

I conclude this chapter by considering another interesting example involving sedimentation, pattern matching, and glaciation. As the glaciers of Europe retreated northward after the last Ice Age, layered sediments accumulated in their marginal lakes or adjacent seas, each layer consisting of coarse material at the bottom, finer material above, and a thin layer of dark mud at the top. These layers, known as "varves" (Swedish: *varv* = layer), were soon recognized as being the product of *annual* depositions of sediment, produced as a result of annual meltings of the glaciers. They were first studied in detail by the Swedish geologist Baron Gerhard De Geer (1858–1943), who initially worked for the Geological Survey of Sweden before being appointed to a chair at Stockholm University, where he later became chancellor and vice-chancellor—and subsequently a prominent Swedish politician.

In 1878, De Geer proposed that varves were analogous to the annual growth rings of trees and thus might be used to establish an absolute chronology for the period of glacial retreat, which took thousands of years. As the lakes gradually shifted northward, there could be no single locality where the total succession could be viewed and analyzed. However, tree rings show seasonal variations, so one can "match" two dead logs with different but overlapping ages by looking for distinctive rings of large or small growth, and this process can be carried through to the trees alive today. Likewise, De Geer and his students compiled a composite chronology for the recent past, correlating one set of varves with another (according to the patterns of annual thicknesses of sediment) from southern to central Sweden. At the keynote address to the International Geological Congress in Stockholm in 1910, his results were laid before his peers in the so-called Swedish time scale (De Geer 1912), covering 12,000 years. Eventually, the scale covered some 15,000 years and was seemingly a wonderful tool for establishing a geological time scale for the late Quaternary, useful also to archeologists.

In itself, De Geer's work was a considerable feat, but he was inclined to push his ideas too far. One problem was that in some years the glaciers might have advanced rather than retreated. More controversially, he made correlations beyond Scandinavia, to North America, East Asia, and even the southern hemisphere, in what were called "teleconnections." Figure 10.6 shows De Geer at work on a varve section in Vermont in 1920, and the caption reveals that he had been there previously in 1891. But he was drawing a long bow in supposing that such strata could be linked one to one with varves in Europe. Certainly, many of his contemporaries thought so in his later career.

Figure 10.6: De Geer examining varves in Vermont, 1920. From De Geer (1940, plate 1). Reproduced by permission of The Royal Swedish Academy of Science.

To back up his idea of teleconnections, De Geer (1927) suggested that the supposed global variations among the varves were due to changes in the heat emitted by the Sun (which was indeed a possibility). Were that the case, then long-distance varve correlations might be possible in principle, even if matching at intercontinental distances was not always practicable. So we had a kind of "solar forcing" hypothesis, but not one that might be accounted for along Milanković lines. (Sunspot cycles, for example, are not linked to changes in the Earth's orbit.)

Thus, cycles entered the story again. But the varve technique did not yield a workable tool accurate to a year. In the 1950s, there arose the question of knowing which layer truly represented the last and youngest varve, and it was shown that De Geer's "zero varve" was out by 84 years because of a correlation error, which was unsatisfactory for a procedure supposed to be precise to a single year. Controversy rumbled on into the 1960s, with the baron's widow continuing to seek to vindicate her husband's results. In the end, "teleconnections" fell out of use as a geochronological tool, being superseded by radiocarbon dating. De Geer's name has not rung as loudly and long as Milanković's.

11

SEQUENCE STRATIGRAPHY AND EUSTASY

The concepts and practices of "sequence stratigraphy" have grown largely from the work of American petroleum geologists post–World War II. They have found wide acceptance in some quarters but rejection or even derision in others. Sequence stratigraphy employs seismic reflection techniques and powerful computers and requires large databases: from surface exposures, well logs, and paleontological records. It is related to cyclothem studies and related matters, and, as with the prewar investigations, it is concerned with the study of recurrent geological structures—and cycles. But it need not be immediately concerned with the ultimate causes of such phenomena, such as the astronomical forces that Milanković studied. Institutionally, sequence stratigraphy has been related to the search for oil as much as a quest for theoretical enlightenment, though the idea that evidence found in sedimentary basins could be accounted for by global and/or astronomical cyclic processes was beguiling, regardless of whether or not oil was found.

We have seen that the concept of "unconformity" was well understood by Hutton (see p. 48). The stratigraphic column was mostly developed in Europe in the nineteenth century, and the major systems (periods)—Cretaceous, Carboniferous, and so on—were established partly according to their lithological features and characteristic fossils and partly by the fact that they were separated by unconformities. But it was a "nice question" whether the European successions and divisions were of worldwide application. Early in the twentieth century, the American stratigrapher Eliot Blackwelder (1880–1969) (1909, 298), who successively held chairs at Wisconsin, Illinois, and Stanford, prepared a diagram showing the areas where, through time, sediments had been deposited in different parts of the United States. It also indicated substantial areas and periods of *non*deposition ("lost intervals/records" or "stratigraphic hiatuses") and was a precursor of what were later called "Wheeler diagrams" (see p. 145).

Blackwelder's unconformities for North America did *not* match those previously established in Europe exactly.

In 1911, Edward Ulrich (1857–1944), paleontologist to the U.S. Survey, published a massive paper, "Revision of the Paleozoic Systems." He was interested in establishing correlations between Europe and the United States but took the view that the American strata might provide better standards. Ulrich was influenced by Chamberlin and Salisbury's ideas on diastrophism (see p. 92), supposing that periodic earth movements led to periodicity in sedimentation both for "grander cycles" and for "subordinate stages." He looked for "natural" divisions of geological time, revealed by widespread unconformities, that separated the packets of sediments that were later termed "sequences." But while he had tectonic activity (diastrophism) as the basis of the major divisions of the stratigraphic column, Ulrich (1911, 401) also said that "the most reliable criteria in determining the periodicity and contemporaneity of diastrophic events, is the level of the sea." Thus, Miall and Miall (2004b, 16) have seen the germs of sequence stratigraphy in Ulrich's paper. Others who envisaged large-scale tectonic events that would lead to periodic changes of sea level and the formation of unconformities between the geological systems, hence providing a basis for global stratigraphies, were the German geologists Hans Stille (1876–1966) (1924) and Erich Haarmann (1882–1945) (1930), the German–American–Chinese geologist and stratigrapher Amadeus Grabau (1870–1946) (1936, 1978), and the Dutchman Johannes Umbgrove (1899–1954) (1947). Stille had an alternation of long periods of relative quiescence, with sedimentation determined by changes in sea level, punctuated by briefer episodes of mountain building or orogeny, the uplifts leading to worldwide marine regressions. Haarmann proposed widespread up-and-down movements (oscillations) of the land surfaces, with concomitant changes in sea level. Grabau, by contrast, had rises and falls of sea level as the primary cause, while Umbgrove thought that pulsatory processes within the Earth caused worldwide eustatic (sea-level) changes. He was the first to publish a general sea-level curve, from the Cambrian to the Pleistocene, though it was based on a chart prepared by Stille, exhibited at Berlin University. These theorists were in the tradition of Suess, envisaging global sea-level changes (see p. 88).

Post–World War II, further problems emerged in the United States in achieving fits with European stratigraphy, and some American geologists began to develop not only their own stages for parts of the stratigraphic column but also a new brand of stratigraphy: the previously mentioned "sequence stratigraphy." These developments occurred in the first instance chiefly at Northwestern University in Illinois and then among petroleum geologists in Oklahoma, Texas, and elsewhere.

Among the innovators was Laurence Sloss (1913–1996), who came to hold a chair at Northwestern University. He studied geology at Stanford and went on to a Ph.D. in paleontology at Chicago. His first appointment was at the School of Mines in Butte, Montana, where he taught paleontology and historical geology. He also worked as a stratigrapher for the State Bureau of

Mines and Geology, which provided facilities for his summer fieldwork. Sloss tried to make sense of Montana's extensive mid-Paleozoic outcrops, which required him to understand the region's subsurface geology and try to form a three-dimensional view of the structure and the regional unconformities. This led to a major paper (Sloss 1950). But before then, he had already moved to Northwestern, where he collaborated with William Krumbein and Edward Dapples to develop ideas on stratigraphy, sedimentology, and tectonics (Sloss, Krumbein, and Dapples 1949).

In his 1950 paper, Sloss produced maps that did not depict outcrops. Rather, they estimated thicknesses, distributions, and erosion surfaces of strata of different facies and different ages for Montana and parts of the adjacent states and north into Canada. It was a major synthesis of the known stratigraphic information for the region (using extensive borehole data), and it began to appear that the surfaces of unconformities did *not* necessarily indicate synchronicity of erosion or unambiguous time lines. There appeared, however, to be large "packages" of sediments—separated from one another by unconformities—that were formed as coherent units, though they were of different ages in their different parts or exhibited different lithologies along given time lines. Such a state of affairs can come about by deposition of sediment in a basin. Coarse sediment is deposited close to the shoreline while at the *same time* fine material is deposited out to sea and at greater depth. Sloss et al. (1949) suggested that there had been distinct *cycles* of deposition, producing four main *sequences* of sediment: the *Sauk* sequence (Lower Cambrian to Lower Ordovician, the oldest), the *Tippecanoe* sequence (Middle Ordovician to Silurian and Lower Devonian), the *Kaskaskia* sequence (Middle/Upper Devonian to Mississippian), and the *Absaroka* sequence (Pennsylvanian to Triassic/Jurassic, the youngest). American tribal names were used for the sequences, just as the Silurian and Ordovician had been named after British tribes. Further sequences were subsequently named after other American tribes. The stages in each cycle involved marine transgression and then regression and were supposedly generated by tectonic agencies. Sloss envisaged tectonic interplay between the areas of geosynclinal activity (the cordilleras to the west) and the stable "craton" of mid-America.[1]

Looking back on his ideas some 40 years after his Montana work, Sloss clarified his terminology as follows:

(1) [T]*he stratigraphy of a cratonic region is divisible into rational and* [pragmatically] *useful packages* [of sediments] *by reference to major regional unconformities.*

(2) [T]*he history of such a region is divisible into tectonically characterized segments representing the times of accumulation of the succession of stratigraphic packages.* (Sloss 1988, 1, 662; emphasis in the original)

Note again that for Sloss the sequences were produced primarily as a result of regional earth movements (tectonics), not global changes in sea level (eustasy).

Ideas about sequences had little impact when first proposed. Sloss (1988, 1662) described his first public presentation of his ideas to the Geological Society of America in 1948 as having "the impact of a failed soap bubble"! But his ideas and techniques were pursued at Northwestern and began to spread from there into the oil industry by the agency of Northwestern students. At the University of Washington, Harry Wheeler (1907–1987), who was professor there from 1948 to 1976 and at times a consultant for Gulf Oil, developed ideas about "time stratigraphy" (Wheeler 1958, 1959) that had much in common with the Northwestern workers' thinking but that also drew on Blackwelder. In fact, Wheeler has been hailed as establishing sequence stratigraphy's theoretical framework, especially with his emphasis on unconformities as indicators of time gaps, for which due allowance should be made in representing the geological history of a region—a point previously made by Blackwelder (and Hutton too, if you like).

As Barrell had emphasized, a vertical sediment log does not give a valid measure of the *time* involved in the deposition, for much material may be removed after deposition or there may be long periods when no sediments are deposited. A log doesn't, in itself, represent a full *history* of a region. To get anything like a history, one has to try to interpret the information recorded in the actually existing rocks so as to "tell a story" about how the sediment pile came to be as it is. And for this purpose, one needed many adjacent logs— and also ones that are widely separated. Then the rocks may be "construed" so that the geological history of the area can be understood. This entails knowledge of the way sediments are deposited *and* eroded through space and time, what may have been happening tectonically in the region concerned, and what may have been happening to sea levels. Such a process, Wheeler emphasized, is necessarily "subjective" to a degree, as compared with the "objective" examination of stratigraphic profiles. He stated (Wheeler 1959, 702) that he was utilizing the Davisian concepts of base leveling and grade in his thinking about sedimentary cycles and his time-stratigraphic diagrams.

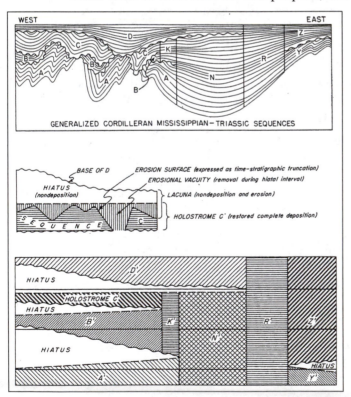

Figure 11.1: Generalized section from the Pacific to the Rockies (northwestern United States) (Wheeler 1958, 1052). AAPG©1958. Reprinted by permission of the AAPG, whose permission is required for further use.

To illustrate his ideas, Wheeler provided a simplified interpretation of the strata from the Pacific coast to the Rockies, along the Fortieth Parallel (Figure 11.1). In this "cartoon," the capital letters represented *sequences*, while the wavy lines represented unconformities. The vertical lines represented (arbitrarily) the boundaries between rocks of different types and ages.

The foregoing diagram could then be construed timewise as shown in Figure 11.2—with time (rather than rock thickness) represented vertically and space (in one dimension) represented horizontally without regard to different rock types. Such diagrams are examples of the previously mentioned Wheeler diagrams.

Figure 11.2 represents an *interpretation* of the depositional events recorded in the rocks shown in Figure 11.1, based largely on the consideration of the various unconformities. Evidently, the region had experienced a series of marine transgressions from the west, followed by regressions and large-scale cycles of deposition and nondeposition.

The unconformable boundaries didn't match the time lines defined by European biostratigraphy. For Wheeler (1958, 1055), the stratigraphic cycle comprised the "entire space–time 'volume' represented by both the surface accumulation (holostrome) and non-deposition (hiatus) interpreted for the interval separating two successive unconformities." The term "holostrome" (which has not come into general use) was proposed (Wheeler 1958, 1056) as "a time-stratigraphic unit embodying the space–time value of a complete (restored) transgressive–regressive depositional succession." It was a sequence *plus* the part of that sequence lost by erosion. (Greek *Holo* = complete; *strom* = layer.) This erosional vacuity was represented in a physical section by a line of unconformity.

Wheeler's idea was essentially sound, but it was difficult to implement, and the interpretative process could be subjec-

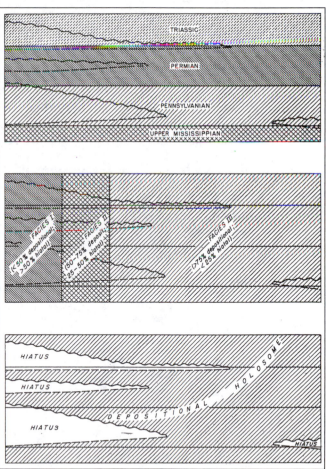

Figure 11.2: Time-stratigraphic diagram, representing the sequences depicted in Figure 11.1 (Wheeler 1958, 1053). The horizontal lines demarcate the Triassic (top/youngest), the Permian, the Pennsylvanian, and the Upper Mississippian (bottom/oldest). AAPG©1958. Reprinted by permission of the AAPG, whose permission is required for further use.

tive. His diagram (Figure 11.1) was only two-dimensional in space, and Figure 11.2 was only one-dimensional in time. To obtain Figure 11.1, one needed much subsurface information, and beyond that considerable "interpretation" was involved. So his "time stratigraphy" did not "catch on" in the 1950s and 1960s. Fossils, which offered approximate information about time lines, continued as the dominant basis for stratigraphy.

Writing in 1963, Sloss emphasized that (for him) sequences were lithostratigraphic units of higher rank than groups or supergroups, being *unconformity-bounded* units. They were "big" entities. Indeed, sequences might extend across North America and even to Britain and Russia (Sloss 1972). As such, they were perhaps more than stratigraphers wanted to deal with and were unwelcome, as they cut across the traditional time boundaries. Nevertheless, using diagrams for his several sequences for strata across North America, rather like Wheeler (Figure 11.1), Sloss (1963) drew a composite time–stratigraphic diagram for the sequences (Figure 11.3), adding two new ones (Zuni and Tejas) to the four proposed in his Montana paper.

But such work was immensely laborious as well as "subjective." It involved the attempted synthesis of vast amounts of scattered data, relating to areas where the subsurface geology was imperfectly known and while geosyncline theory still held sway in the United States. So the underlying tectonic theory was, we now think, suspect. Progress was made, however, particularly through the contributions of petroleum geologists and through the help provided by their seismic studies in imaging subsurface structures. From such work, there eventually emerged the field now termed "sequence stratigraphy."

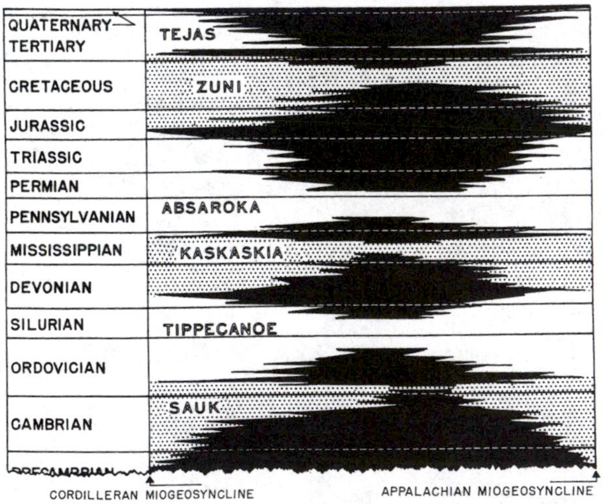

In 1956, one of the Northwestern students, Peter Vail (b. 1930), who had studied under Sloss, Krumbein, and Dapples, moved to Carter Oil Company in Tulsa, Oklahoma, relocating in 1965 to Houston, Texas, in the Esso Production Research Company (later the ExxonMobil Upstream Research Company), which subsequently became part of the Exxon Group. He was taken on to evaluate, develop, and apply new mapping techniques to assist oil exploration, working with Robert Mitchum, another Northwestern man, and also with Krumbein (who was a consultant for Carter Oil while also teaching at Northwestern). Vail (and others) visited Carter's regional offices and collected well-log data and regional seismic sections. The develop-

Figure 11.3: Time-stratigraphic relationships for sequences of the North American craton (Sloss 1963, 110). The black areas represented nondepositional hiatuses; white or stippled areas represented the occurrence of sediments in space and time. (The stippling was added to give clarity to the diagram and didn't denote rocks that were different from those in the white areas.) Reproduced by courtesy of the Geological Society of America.

ments in seismology were such that geologists were able to image subsurface stratification patterns by analyzing the reflections (from the surfaces of sedimentary layers) of waves generated at ground level by small explosions—rather like bats identifying objects in their vicinity by "analyzing" the reflections of sound waves that they emit while in flight.

Following Vail's career at Exxon, he was appointed to the W. Maurice Ewing Chair of Oceanography at Rice University, Houston, in 1986. When he was awarded the Geological Society of America's Penrose Medal in 2003, his former colleague Robert Mitchum said in the citation,

> When Peter Vail introduced the concepts and applications of sequence stratigraphy, the effects on stratigraphic geology and seismic interpretation were comparable to those of plate tectonics on structural geology. Pete's ideas on the unifying paradigm of eustatic cycles are probably as close to an original concept as most of us are privileged to see. Pete's worldwide experience with Exxon's exploration groups honed the concept into an immensely practical tool for hydrocarbon exploration and stratigraphic studies in general. (Mitchum 2003)

Evidently, sequence stratigraphy made a big impression in some quarters!

Vail's correlations between well-log data deployed the emerging computer power, the value of which he had learned from Krumbein at Northwestern. Then, following a suggestion from his colleague Charles Campbell, he came to the view that time lines could be determined from physical stratigraphy (correlations between bedding surfaces or between physical/stratal patterns) as well as from biostratigraphy. For example, if sediments are deposited on a sloping submarine shelf, an inclined bedding plane will approximate to a time line that can be imaged by seismic techniques. In following a bedding plane, the rock type may change from, for example, coarse to fine as sediment is deposited synchronously at increasing distances from a shoreline and at increasing depth. So bedding planes, not facies changes, were manifestations of time lines.

Vail and his group reached this conclusion by studying the correlation of marker beds in well logs from which he could infer a series of stratal patterns that represented "offlap," "downlap," and "onlap" (see Glossary). On showing these results to a seismic interpreter, Paul Tucker, Vail was told that such patterns were well known in seismic profiles. So he began to interest himself in seismic techniques.

To learn such techniques, Vail partly moved into geophysics, and during the late 1950s and 1960s, he and his colleagues worked on well-log correlations, with data from increasingly distant parts of the world, looking for indications of unconformities and possible long-distance correlations. The seismic techniques made possible the study of the complex "architectures" or structures of sediments in sedimentary basins (such as the North Sea, where much work was conducted at that time examining that basin's oilfield), and efforts were made to establish temporal and stratal correlations of sequences (packets of sediments) in different basins. A worldwide database of seismic sections, well records, and biostratigraphic data was assembled within the Exxon Group. Vail looked for evidence of *global* stratigraphic correlations, with indications of concurrent *global* onlaps and offlaps. He has recorded (Vail 1992, 87) that in 1959 he attempted to draw a global sea-level cycle chart for such changes from Cambrian to Recent and gave a talk on his results at a company forum in 1963. In 1966, he made his ideas more public with a talk with Robert Wilbur at a meeting of the American Association of Petroleum Geologists. In the published abstract (Vail and Wilbur 1966), it was suggested that the sediments were deposited relatively rapidly at the beginning of the onlap part of a cycle and died out toward the end of the onlap. In the next cycle, the initial onlap was again rapid but occurred closer to the center of the sediment basin. The changes in relative levels of land and sea were thought to be caused by global changes of ocean levels due to a combination of tectonic processes and climatic changes. The authors suggested that they had identified *worldwide* unconformities that corresponded with some of the major subdivisions of the stratigraphic column (e.g., between Silurian and Devonian).

Vail's onlap–offlap chart (not then published, but see Figure 11.4) appeared to show a "sawtooth" pattern through time, as (it seemed) relative sea levels rose—and then fell quite suddenly. Thus, it appeared, there was evidence for global eustasy governing sedimentation—perhaps controlled by general climatic changes, with the accumulation or melting of ice around the North Pole and the South Pole. I don't know whether Vail was thinking about Milanković cycles at that fairly early stage of his work, but it seems likely that his thoughts turned in that direction. Certainly his abstract with Wilbur suggested that changes in sea level could be due to ice accumulation or melting.

Although Vail was giving out some information about his work to the scientific community, much of it remained with Exxon ("commercial in confidence"). He applied his techniques to the North Sea basin, for example, and "it contributed to a number of blocks being leased by Shell/Esso in the central North Sea" (Vail 1992, 87). We can understand, therefore, why the Exxon techniques were not immediately broadcast to the world, and only abstracts of talks were published. Eventually, Vail's group revealed their results in 1977 in a paper in which the so-called Vail curve first appeared (Vail et al. 1977: Vail, Mitchum, Todd, and Widmier [Exxon Production Research Co., Houston], Thompson [New Mexico Bureau of Mines and Mineral Resources], Sangree [Esso, UK], Bubb [Exxon Production Malaysia], and Hatlelid [Imperial Oil Ltd, Calgary], see Figure 11.4). It was the published version of a presentation before a Geological Society of America meeting in 1975. Various more detailed curves were also published in the 1977 volume.

Vail also indicated how his sawtooth curves could be inferred from the field data, taking an example from northwestern Africa (Figure 11.5). First, a stratigraphic profile was prepared on the basis of seismic and well-log data and so on (Figure 11.5a). (Note the theoretical interpretations of the information inherent in the key to this diagram. A, B, C, D, and E represented distinguishable "sequences.") From this information, a chronostratigraphic chart (Wheeler diagram) was prepared (Figure 11.5b). To accomplish this, the unconformities apparent in Figure 11.5a were plotted onto the time diagram of Figure 11.5b, which showed the sediments deposited at different points along the profile

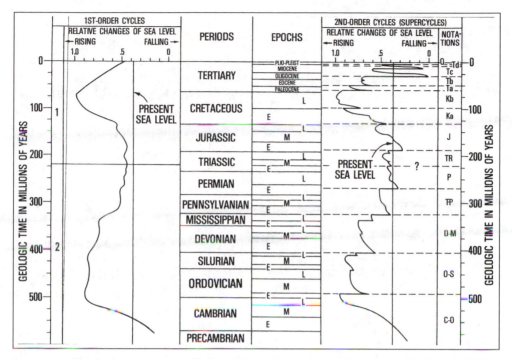

Figure 11.4: The Vail curve, as unveiled in 1977, with two levels of resolution (Vail *et al.* 1977a, 84). E = Early; M = Middle; L = Late; C = Cambrian; O = Ordovician; S = Silurian; D = Devonian; M = Mississippian; PP = Pennsylvanian–Permian; J = Jurassic; TR = Triassic; Ka,b = divisions of the K[C]retaceous; Ta,b,c,d = divisions of the Tertiary (all somewhat approximate equivalences). AAPG©1977. Reprinted by permission of the American Association of Petroleum Geologists, whose permission is required for further use.

at different times and also "spaces" representing nondeposition. Times were based on paleontological and radiometric data and so on. The seismic reflectors were assumed to serve as time markers, but the planes of unconformity could have been tilted by tectonic activity. A sawtooth sea-level curve (Figure 11.5c) was then produced by interpretation of Figure 11.5b, assuming slow sea-level rises and comparatively rapid falls (manifested in the horizontal portions of the sea-level/time graph). The procedure was then extended to yield the global chart of Figure 11.4 by "averaging" information acquired in different parts of the globe (Figure 11.6). (However, the global diagram at the right of this diagram does not appear to be a plausible "average" of the data from Australia, the North Sea, northwestern Africa, and California.)

The work of Vail and his coauthors was evidently intended to apply on a global scale, though the term "sequence" was used by the oil workers to refer to packages of sediments on a much smaller scale than those that Sloss contemplated in his original investigations. According to Sloss (1988, 1664), the Vail curve (Figure 11.4) and others like it in the Payton (1977) volume and subsequent publications aroused both interest and controversy. People complained that the data from which the curves were compiled were unavailable to the scientific community as a whole, higher resolution for the Cretaceous than that shown in Figure 11.4 was available to Exxon but was not published, and Vail et al. had made unwarranted assumptions that the results could largely be ascribed to sea-level

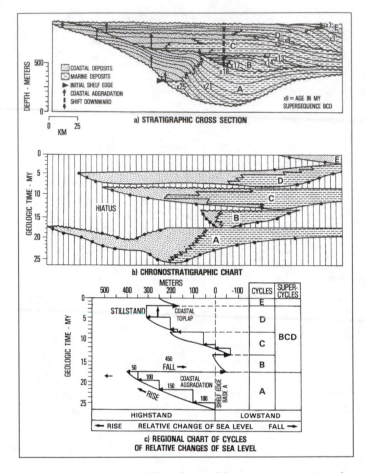

a) STRATIGRAPHIC CROSS SECTION

b) CHRONOSTRATIGRAPHIC CHART

c) REGIONAL CHART OF CYCLES
OF RELATIVE CHANGES OF SEA LEVEL

Figure 11.5: Conversion of data obtained from seismic stratigraphy into part of a Vail curve (Vail *et al.* 1977a, 78). AAPG©1977. Reprinted by permission of the American Association of Petroleum Geologists, whose permission is required for further use. B, C, and D were thought of as "supersequences" (later called "megacycles").

changes per se without regard to tectonic factors. In addition, it seemed to be assumed, without warrant, that rises in sea level were gradual, whereas falls were sudden (giving the sawtooth appearance to the Exxon diagrams).

However, in subsequent years, the sawtooth curves were not ascribed specifically to changes of sea level. Instead, there was talk of relative changes of coastal onlap. Sloss (1988, 1664) also pointed out that "the 1977 curves represent[ed] two decades of analysis of data . . . including thousands of kilometers of seismic lines, hundreds of subsurface records, and untold man-hours of biostratigraphic work." On the other hand, as indicated previously, much of the work was "proprietary." Later sea-level diagrams (e.g., Hallam 2004, 90) show more "rounded" curves rather than the sawtoothed appearance of the earlier illustrations, and Miall (1990, 464) has said that the "famous sawtooth coastal

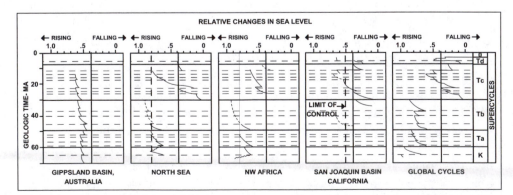

Figure 11.6: The "averaging" of sea-level curves from four different regions to yield a global sea-level curve, right-hand drawing (Vail *et al.* 1977b, 90). AAPG©1977. Reprinted by permission of the American Association of Petroleum Geologists, whose permission is required for further use.

onlap curve . . . has now been retired." A relatively recent "smoothed" Vail curve appears in, for example, Duval, Cramez, and Vail (1998, 46).

It should be noted that Vail curves could be displayed with greater or less precision (two levels of detail appear in Figure 11.4), and Vail and his coauthors proposed a hierarchy of processes, revealed in the curves according to their degree of precision. In his first main publication on the topic, Vail et al. (1977, 86) referred to cycles of three "orders of magnitude." There were only two of the longest, or "first-order," ones: Precambrian to Early Triassic (about 300 million years ago [Ma]) and Middle Triassic to the present (about 225 Ma). There were supposedly 14 "second-order" cycles (10 to 80 Ma), and more than 80 "third-order" cycles (1 to 10 Ma), which became apparent as more data were supplied to the compilation of the diagrams. It was as if "nature" were being examined with ever-increasing magnification.

In subsequent work, six orders of cycles were proposed: (1) 50+ Ma, (2) 3 to 50 Ma, (3) 0.5 to 3 Ma, (4) 0.08 to 0.5 Ma, (5) 0.03 to 0.08 Ma, and (6) 0.01 to 0.03 (Vail et al. 1991, 619). The first-order cycle was thought to be primarily tectonic in origin, whereas the second to fifth orders were said to be glacio-eustatic cycles (Vail et al. 1991, 643). They would be manifested as sedimentary records of changes in sea level, occurring as small "waves" (or cycles) of changes imposed on larger "waves" imposed on yet larger "waves" (Einsele and Ricken 1991, 612) and so on in a way reminiscent of what Barrell had envisaged back in 1917. Evidently, Vail's theory "evolved" after its first public enunciation. It was claimed that the different cycles could be recognized by their characteristic "signatures" (specific features) in the sedimentary record.

In the discussions following the first publication of the Exxon Group's ideas in 1977, there was much consideration of the "causes" of the cycles and in particular the role that tectonic factors did or did not play in the picture. The Group was accused of giving undue prominence to global eustasy caused by climatic changes. This is an overstatement, as Vail et al. certainly did not exclude tectonic factors (thought of by Suess and other European geologists to be responsible for global eustasy). On the other hand, we find that even for third-order cycles, which were attributed by Vail et al. (1991, 619) to folding, faulting, and diapirism (i.e., the piercing or rupturing of uplifted rocks by upwardly mobile low-density materials such as salt deposits or by upwardly mobile Earth-core material), these supposedly only affected the *amplitude* of sea-level changes, not their *timing* (Vail et al. 1991, 619).

Others have preferred five orders of cycles, as has been conveniently tabulated by Andrew Miall (1995, 10), professor at the University of Toronto, with tectonic factors being involved for all orders (see Table 11.1).

It may not be obvious from the foregoing how sequence stratigraphy is useful to oil geologists. But the main point for present purposes was, perhaps, that stratigraphers began to develop techniques that enabled them to understand the *architecture* of sedimentary basins in terms of sedimentological deposition theory, time stratigraphy, cycles of sedimentation, and (perhaps) ideas about eustasy and global climatic changes. (Sequence stratigraphy does not *require* assumptions about eustasy.)

Table 11.1: Stratigraphical cycles according to sequence stratigraphy theorists.

Type	Terminology	Duration (in Ma)	Probable causes
1st order		200–400	Major cycles caused by break-up of super-continents
2nd order	Supercycles (Vail 1977); sequences (Sloss 1963)	10–100	Eustatic cycles caused by global mid-ocean spreading. Regional extensional down-warp and crustal loading.
3rd order	Mesothems (Ramsbottom 1979); megacyclothems (Heckel 1986)	1–10	Regional cycles caused by intra-plate stresses. Probably not of global extent.
4th order	Cyclothems (Wanless and Weller 1932); major cycles (Heckel, 1986)	0.2–0.5	Milanković glacio-eustatic cycles; astronomical forcing. Regional cycles caused by crustal flexure, especially in foreland basins.
5th order	Minor cycles (Heckel 1986)	0.01–0.2	Milanković glacio-eustatic cycles; astronomical forcing. Regional cycles caused by crustal flexure, especially in foreland basins.

Sequence stratigraphy has been described in many publications, and one by Neal, Risch, and Vail (1993) may help explain its relevance to oil geologists. The Exxon Group envisaged the successive deposition of five stages of sedimentation in a basin in response to rising sea levels:

1. Sand-rich deposits formed in basin while sea level is at its lowest (conventionally represented in yellow by the Exxon school).
2. Sands and shales deposited in fans on the continental slope (conventionally represented in brown, with slumps represented in purple). It is the part of the continental margin, inclined at about three to six degrees, between the continental shelf and the continental rise. The continental rise is the very gently inclined region between the continental slope and the plains of the ocean deeps. It can have submarine channels with levees that meander across the fans, and slumps are common (conventionally represented in brown).
3. Wedges of sediment deposited in the basin, with silts and shales formed seaward and some sands nearer the shore (conventionally represented in red).
4. Sandy material formed as beaches and sandbars as sea level rises relatively rapidly (conventionally represented in green).
5. Sediments building out and covering (4), as the sea level rises more slowly near the close of a period of sea-level rise: the seaward sediments are sandy, while the landward material can be the muds of estuaries and marshlands (conventionally represented in orange).

The foregoing together make up a sequence. Any fall in sea level will then produce an upper erosion surface or unconformity. The five sets of sediments will also lie unconformably on the previous sequence onto which it has been deposited. All five components are not invariably present. But, when present, the different components can be recognized in seismic profiles according to their characteristic sedimentary structures and features. In addition, each sequence component has its characteristic signals for gamma radiation, resistivity, and "spontaneous potential" (the electrical potential difference between the drilling mud and the water in a formation, indicative of salinity), which can be identified by geophysical instruments lowered into a borehole or by examination of core samples. The profile of a typical sequence is shown in Figure 11.7 (upper drawing), with a corresponding Wheeler diagram (lower drawing).

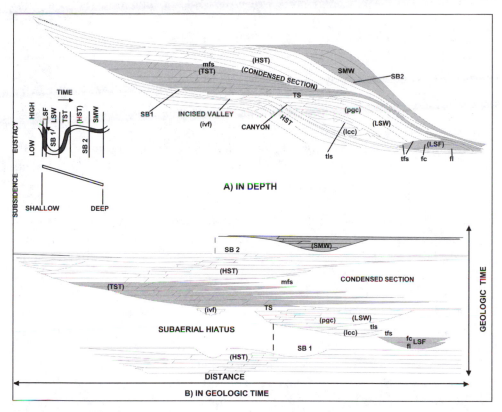

Figure 11.7: Profile of a typical sequence and a corresponding Wheeler diagram (Haq *et al.* 1987, 1157). Reprinted, with permission, from Haq et al., "Chronology of Fluctuating Sea Levels since the Triassic," *Science*, 1987, 1157. © 1987 AAAS.
Key:

Surfaces
SB = Sequence boundary
DLS = Downlap surfaces
mfs = maximum flooding surface
tfs = top fan surface
tls = top leveed channel surface
TS =   Transgressive surface
(first flooding surface above maximum regression)
Systems tracts
HST =   Highstand systems tract (5)
TST =   Transgressive systems tract (4)
LSW = Lowstand wedge systems tract (3 + 2)
ivf = incised valley fill

pgc = prograding complex (3)
lcc = leveed channel complex (2)
LSF =   Lowstand fan systems tract (1)
fc = fan channels (1)
fl = fan lobes (1)
SMW = Shelf margin wedge systems tract (beginning of a new sequence as sediments are deposited offshore prior to a fall of sea level and the repetition of the processes of deposition)
For a three-dimensional colored representation of the stages of the formation of a sequence, see: Neal et al. (1993) at: www.oilfield.slbcom/media/resources/oilfieldreview/ors93/0193/p51_62.pdf.

Sequences such as those shown in Figure 11.7a may be found stacked on top of one another in big sedimentary basins, such as the Gulf of Mexico. Large sandy, basin-floor sand deposits, with impervious sealing-shales above and below them, are likely sites for oil accumulation (as are dome-like structures). So studies of seismic profiles and well-log information may guide decisions as to where to drill for oil. Thus, Vail's work on the architecture of sedimentary basins had economic implications regardless of whether the idea of a global curve for sea-level changes was justified, and the ideas about eustasy and sea-level curves probably didn't enhance oil companies' bottom lines a great deal. So while sequence stratigraphy per se does not necessarily help much in *finding* oil, it can help sort things out after it has been discovered.

However, while understanding the architecture of sedimentary basins has been important, as have been the Exxon Group's techniques and databases, the ideas about global sea-level curves have raised many hackles (though the Exxon Group has asserted "new paradigm" status for their procedures and results). We recall that Vail saw it as perhaps his major insight that the stratal boundaries revealed by seismic investigation or otherwise represented *time lines*. Traditionally, time lines in geochronology have been provided by fossils, radioactive determinations of rock ages, the study of volcanic ash deposits, or, as we will see in Chapters 12 and 13, geomagnetic or geochemical investigations (determination of isotope ratios). One may also use "Milankovićian" calculations for relatively young rocks to calculate the Earth's former insolation and hence estimate what past climates have been. Hence, sea levels may be estimated. So, assuming the validity of the astronomical calculations, a rock unit (as part of a sedimentary cycle) might be *dated* by considering whether it represented a high-sea-level deposit or a low-sea-level item. This gave a whole new dimension to the concept and scope of cyclicity and an expanded domain of applicability.

Of course, the early sea-level curve of Vail et al. (1977) was not used as an independent criterion of geological time. But later, more precise curves, based on more data and linked to geomagnetics and biostratigraphy, such as the frequently cited figure of Haq, Hardenbol, and Vail (1987; see Figure 11.7a), all of the Exxon Group, have began to appear as *arbiters* of geological time. This has led to objections. For example, while accepting that the "sequence-architecture model" is well established and useful, Miall (1994, 1995; Miall and Miall, 2001) has been concerned that the sequence boundaries have been represented on Exxon diagrams with separations in time *less* than the error margins of the data used to prepare the figures. It is also said that too little weight has been given by the Exxon school to tectonic factors and that for the higher-order cycles these have been largely ignored. There was also the issue, previously mentioned, of the repeated slow sea-level rises and sudden falls (giving the sawtooth appearance of the early Vail curves). This would seem difficult to relate to glacio-eustatic control of the phenomena (or anything else). Nevertheless, many have expressed satisfaction with the application of the sequence-stratigraphic model to the phenomena of cyclothems.

Another persistent objection has been that sequence stratigraphers may argue in circles (not cycles!), for "[t]he global-eustasy hypothesis seems to have begun to work a powerful influence on the selection and collation of additional data" (Miall and Miall 2001, 328). These authors claim that the sequence-stratigraphy workers suppose that their findings provide fresh evidence in favor of global changes *because* they correlate with the standard global charts of Vail et al., and hence the correlations can supposedly be used to extract or isolate the eustatic components of the causality. Miall and Miall attribute such practices to practitioners adhering to the "paradigm" of the global eustasy model. The critics draw attention to the counterparadigm—the "complexity paradigm"—of authors such as John Dewey and Walter Pitman, who have stated that they "cannot conceive of a scenario that would allow the synchronous global-eustatic frequency and amplitude implied by third-order cycles" (Dewey and Pitman 1998, 13). They prefer the multicausal hypothesis that sea-level changes can be due to several causes that may be operating simultaneously and at different rates over different ranges of space and time rather than attributing all sea-level changes to changes in ocean basin volumes for the low-frequency cycles and glacio-eustasy for the high-frequency changes.

Thus, the opponents have fronted off at one another. Tempers were not improved when Miall (1992, 789) published a diagram in which the Exxon chart of 1988 for 40 Cretaceous sequence boundaries plotted on a vertical time axis were compared with four other columns of geological data that correlated with the Exxon data with a success fit rate of 77 percent or better for time differences of up to 1 million years—a result that might seem satisfactory to most geologists involved in the imprecise practice of correlation. The catch was, however, that the four sections were constructed from random number sequences!

Recently, the husband-and-wife team of Andrew and Charlene Miall (the latter a sociologist/social psychologist at McMaster University, where she does work in the field of "social constructivism") have published interesting papers on the work of the Exxon Group and have mounted a substantial critique of its ideas and practices (Miall 1992; Miall and Miall 2001, 2002, 2004a, 2004b). In so doing, they have done historical research, interviewing workers from the Exxon Group, and reporting (without attribution) some of their statements made in confidence. Their findings are rather extraordinary.

Within Exxon, it appears, there were considerable tensions, as the creative and well-resourced Group's model was developed in the 1960s and 1970s. There was rivalry between the company's workers in different divisions. The geophysicists who did the seismic work felt themselves relegated to the unaccustomed role of data collectors and processors and rejected the view of Vail and his geologist colleagues that seismic reflectors followed time lines. Vail's own position in his company was ambiguous, lying somewhere between the areas of geology and geophysics. When his seismic stratigraphy group was restarted in 1974, by the authority of James Coffman, an Exxon vice president and director, it was done against the objections of most of the regional and project managers.

The first substantial publication of the work of the Exxon Group (Payton 1977), issued by Coffman's authority (but including no data less than three years old), was unusual for a theoretically important book in that it appeared without refereeing external to Exxon. (It couldn't be, as the academic community didn't have access to the "commercial in confidence" data, and the outsiders had no means of judging which observations had been included and which had been arbitrarily omitted.) This situation still remained when the prestigious journal *Science* published the work of Haq et al. (1987), which set out more detailed curves for the Triassic and younger strata. By this time, however, geologists were climbing aboard the Exxon bandwagon, even though their data still had to be taken on trust (but they were finding oil successfully). The "bandwagoners" well knew of Exxon's massive facilities and database. These factors, plus the publication in *Science*, persuaded many academic geologists that the work must be sound. Other oil companies began to deploy the Exxon methods. Moreover, the doubters within the Exxon camp began to be convinced of the worth of their procedures and findings in the light of the favorable reception of their work. So the bandwagon rolled, and the Exxon people, encouraged, began to expand their sequence-stratigraphy program. (This paragraph is based on Miall and Miall 2002.)

In an influential book by the French sociologist of science, Bruno Latour (1987), there is an insightful discussion of what he calls "black boxes." If you undertake an experiment, you will want your apparatus—and its use—to be accepted as reliable by your peers and not in need of question, examination, or criticism. If opponents disagree with the results obtained with your apparatus, they will have to come and visit you and see your apparatus, how it works, and how you use it. Or they must "problematize" the box, making it "gray" or partly transparent—for example, by questioning the validity of the standards it employs, the purity of the chemical agents used in its operation, the precautions used to allow for other errors in the experimentation, and so on. To bring down their opponent's work, they must build a *better* piece of apparatus and perform a better experiment. Others in the relevant community will judge which piece of research or the work of which group is to be regarded most favorably and built on.

One can apply such ideas to databases. The Exxon database was in effect a black box as far as the rest of the geological community was concerned. One had moved away from the ideal of "pure science" into the realm of "technoscience," where proprietary considerations were prominent. No one could "replicate" the Exxon work as such. The data were unavailable, and the giant oil company's resources were inaccessible to other players in the field (though, as mentioned previously, other oil companies did begin a catch-up process—but for the purpose of finding oil, not to undermine the Exxon Group's methodology). It was not easy to "whiten" the black box of the Exxon database directly or to render it transparent. (Another interesting case for a geological database has been described by Le Grand and Glen 1993). So it took time for the sequence sequence-stratigraphic model to come under proper scrutiny, and Miall and Miall (2002, 323) report that when Exxon people were with other geologists in Britain in 1984, they tended to "prejudge" the interpretation of the field evidence in the light of their own scheme rather than testing that against the field observations. There were some heated arguments.

This again raises the question of possible circular reasoning in sequence stratigraphy and the global eustasy paradigm. In a paper by Haq, Hardenbol, and Vail (1988), the authors specified cycles of sea-level changes based on an array of seismic, well-log, and outcrop data, with radiometric, biostratigraphic, and magnetostratigraphic (see Chapter 12) information incorporated. Thus, a new generation of improved cycle charts was offered. The primary seismic data were not placed in view, but sedimentological and paleontological data were provided in the sequence charts, and much geomagnetic information was given. Then, "[o]nce this [sea-level] framework ha[d] been constructed, the depositional sequences from sections round the world, interpreted in response to sea-level fluctuations, c[ould] be tied into the chronostratigraphy" (71). That is, new data were construed *in terms of* the previously established sea-level curves. This tendency has persisted in later publications of the "global cycle chart school" (e.g., De Graciansky, Hardenbol, Jacquin, and Vail 1998). But the claimed precision of the sequence boundaries in this publication was greater than that which chronostratigraphic resolution could offer for the strata in question (a previously mentioned problem). Tectonic effects were merely taken to enhance or subdue the eustatic changes rather than having an independent role. Where "new" sequence boundaries were found in parts of the world that had not been used in the construction of the original sea-level curves, they were simply added to the original number, with the result that there was a proliferation of "events" in the curves and their average spacing was reduced, thus conveniently bringing some of the cycles into the "Milanković band" (10,000 to 500,000 years BP), where the eustatic phenomena could supposedly find a satisfactory astronomical/physical explanation.

The foregoing and other criticisms have been made in Miall and Miall (2001, 2004a) and by others such as William Dickinson (2003) of the University of Arizona, who reconsiders the fundamental postulate of Exxon sequence stratigraphy, namely, that seismic reflectors are chronostratigraphic horizons. While accepting that this may be the case for practical purposes for specific strata or rock layers, he rejects the idea that it holds also for the boundaries of unconformities, for these can readily be diachronous, as, for example, when a sea level rises and water slowly encroaches on the land. The idea of a synchronous global stratigraphy is, Dickinson suggests, incapable of independent test and has no known causal mechanism other than episodes of glaciation and deglaciation. He therefore places it in the category of geomyths. There I shall leave it: it is not the historian of science's role to adjudicate contemporary controversies. But as far as I'm aware at the time of this writing, the criticisms by the Mialls and others have not yet been refuted in refereed publications.

As a codicil to this chapter, I note a late paper by Sloss (1991), whose work launched Vail's career in the direction of sequence stratigraphy. Sloss was evidently concerned that the Exxon Group was giving insufficient attention to tectonic factors in its explanations of the several orders of cyclicity. His critique lies beyond the scope of the present book. But the *title* is noteworthy:

"The Tectonic Factor in Sea Level Change: A Counter*vail*ing View"! (I thank Professor Miall for drawing my attention to this evidently intentional pun.)

NOTE

1. The word "craton" (Greek *kraton* = strength or power) derives from the term "kratogen" of the Austrian geologist Leopold Kober, who in 1921 divided the Earth's crust into "orogens" and "kratogens," the former being mountain areas developed from geosynclinal belts squeezed between stable "forelands" or "kratogens." Hans Stille changed the term to "kraton" in 1936, and it was further changed to "craton" by the American Marshall Kay in 1944.

12

THE GEOMAGNETIC TIME SCALE AND RELATED MATTERS

Looking back to Chapter 8, we recall that an essential part of the "plate tectonics revolution" was the recognition of "magnetic stripes" in the basaltic rocks forming the ocean floors. These represented the Earth's magnetic field directions at the time the basalts were erupted onto the ocean floors from the mid-oceanic ridges and then moved "sidewise" as new oceanic crust formed. The further a "stripe" lay from a ridge, the *older* it was.

This chapter takes up this theme but extends it by discussing the magnetic reversals that may be found in *vertical* sections of the crust. Since the reversals are global and occur irregularly and almost instantaneously (geologically speaking), they can provide an additional component in the armory of the stratigrapher, to be considered along with paleontological evidence and the supposed sea-level changes and Milanković cycles—for the reversals provide *global* "time markers." So we find that geomagnetic reversals are commonly shown on charts depicting geological time scales, especially for the Tertiary (Cenozoic) strata (e.g., Gradstein, Ogg, and Smith 2004). The changes in magnetic polarity are analogous to two-way switches, so that the resultant magnetic pattern in the rocks somewhat resembles a bar code. If we accept this analogy, it's obviously imperative that all geomagnetic reversals be detected and included in the vertical table. If any are missing, the whole scale could be useless for geochronological purposes, as is a bar code if one or more of its black lines are missing because of a "printing error."

Before embarking on a historical examination of the topic, some words are in order concerning the origin of the Earth's magnetic field and its reversals. As is well known, there are magnetic poles in the Arctic and Antarctic regions, but they are not fixed, and a line drawn between them doesn't generally coincide with the Earth's axis of rotation. The magnetic poles move slowly and somewhat irregularly relative to the geographical poles. Moreover, a line drawn from one magnetic pole to the other would not generally pass through the Earth's center, and the wanderings of the magnetic poles seem to take place independently of one another. There are also some "subsidiary poles." Such phenomena cannot be accounted for by reference to changes in the Earth's rotation, which is almost constant, nor can there be some kind of "bar magnet" within the Earth.

An explanation of geomagnetism has long been sought in some kind of self-sustaining dynamo in the Earth: there are circulating electric currents that generate inconstant magnetic fields, and the changing magnetic fields generate currents. But the dynamo hypothesis is difficult to model mathematically so as to allow prediction of the known geomagnetic effects. It lies beyond the scope of this book to give a proper account of geomagnetic theory, but the following qualitative remarks may be of assistance.

To expand information in Chapter 8, the Earth (radius about 6,400 kilometers) is thought to consist of a solid *inner core* (at a depth of about 5,200 kilometers), a fluid *outer core* (at a depth of about 2,900 kilometers), and an "almost solid" (i.e., plastic) *mantle* of rocky material, which can slowly flow like pitch and occupies the region between the thin, solid, and rigid outer *crust* on which we stand. This crust is up to about 30 kilometers thick. The fluid part of the core is thought to consist chiefly of iron. There is a significantly more plastic *asthenosphere* in the upper mantle that allows the lateral movement of the crustal "plates." The so-called *lithosphere*, about 80 kilometers thick, lies above the asthenosphere and is made up of the crust and part of the upper mantle. Such information is derived principally from seismology.

Given the existence of radioactive heat-generating minerals within the Earth and perhaps also heat left over from that which was generated when the Earth first formed, supposedly by the accretion of "loose" material circulating around the Sun, one can envisage the existence of convection "cells" within the fluid core. These cells are supposedly the "engine" that drives the dynamo that produces the Earth's magnetic field.

Imagine magnetic lines of force passing through a stationary fluid core of high electrical conductivity within the Earth. In such a situation, the field lines will be straight (in a small volume of material). If, however, the material is being sheared as a result of convection currents within the core, the lines of force will be moved or distorted in some manner, and the moving lines of force (equivalent to a changing magnetic field) will generate a current that will produce changes to the magnetic field. Thus, we have a self-sustaining dynamo. The source of the initial "start-up" field is uncertain, but it could be, for example, a product of the Sun's magnetic field or it could be due to electric currents arising from the different electrochemical potentials of the different elements in the core and mantle. Whatever: a geomagnetic field could arise from the combined effect of circuits of electric current within the core, resulting from convection and shearing (the Earth is spinning) and also probably from sources external to the Earth.

As a thought experiment, one can imagine two diametrically opposed convection cells, of approximately equal magnitude, generating opposed magnetic fields. The combined effect would produce a combined magnetic polarity, directed one way or the other at any given time, according to which convection current is producing the greater effect. The overall field strength will be small if the two opposed cells are generating approximately equal magnetic fields. However, the convection current that is "winning" (generating the stronger magnetic field) can change over time.

Next imagine a number of convection cells, all varying over time, since the phenomena are unsteady as a result of the disordered nature of the thermal convection currents. Such a state of affairs would naturally cause many minor variations in the Earth's field, such as are in fact observed. In addition, the combined effect of the various cells could be such as to either decrease or increase the overall field strength. If it is decreasing, it might from time to time arrive at zero field strength, and were that to happen, it would be a matter of chance whether, as the field later increases, it does so with the same polarity as before or with reversed polarity. Thus, there need be no mystery about the geological record revealing irregular geomagnetic reversals.

It should be emphasized, however, that the convection cells must have a more complex structure than that hypothesized previously, and with the Earth's rotation they will tend to have some kind of spiral form (arising from the so-called Coriolis force). Fortunately, however, for our present purposes the *explanation* of the formation of the geomagnetic field and its reversals need not concern us further. It is sufficient that there *is* a field that can reverse from time to time—globally.

The fact that rocks such as basalts could be magnetized by the Earth's magnetic field was remarked (*en passant*) in the nineteenth century by the French mineralogist, petrologist, and mining engineer Achille Delesse (1817–1881) (1849, 200). Then a French geophysicist, Bernard Brunhes (1867–1910), and Pierre David found that the iron-bearing materials in some of the sediments baked by lava flows in the Puy-de-Dôme region of central France were magnetized, as were the lavas that had caused the baking, and their magnetization was aligned with the Earth's field. More surprisingly, they found that the rocks possessed a magnetization that was *opposite* to the Earth's present field (Brunhes 1906; Brunhes and David 1901, 1903). So, they concluded, the Earth's field must at some time have been in the opposite direction to what it is at present. But at the time, this seemed so unlikely that the discovery was largely ignored.

Two decades later, the French work gained support from the Japanese geophysicist Motonori Matuyama (1884–1958) of Kyoto University, following his studies of lava flows in Japan, Korea, and Manchuria, which led him to think that the Earth's field had reversed itself during the Pleistocene and, from results reported by Paul-Louis Mercanton, in earlier epochs (Permo-Carboniferous and Miocene) as well (Matuyama 1929). But the idea that there were global geomagnetic reversals was queried for a time, as some minerals (such as pyrrhotite) acquire a *reverse* field when formed in a *normal* field. So the idea of global reversals didn't catch on, and it was not until the 1950s and

1960s that sufficient evidence had accumulated to convince geologists that the Earth's field reversed from time to time, though not according to any regular pattern in time (Cox, Doell, and Dalrymple 1963). The important point to demonstrate was that magnetic reversals occurred *globally* and could be dated to show that they occurred simultaneously in different parts of the world. Dating was possible since lavas could be dated accurately by measuring the proportions of potassium and argon in a rock sample (One of the potassium isotopes, ^{40}K, decays spontaneously and at a known rate into the inert gas argon, ^{40}Ar. So, proportionately, the more ^{40}Ar trapped within a sample, the older is the rock, and thus it can be dated.) So if a lava lying immediately above the horizon of a magnetic reversal in a rock suite is dated and one immediately below, then the date of the reversal may be pinpointed. The reversals appeared to be rapid in terms of geological time: perhaps only 10,000 years. Other radiometric dating procedures can also be used, appropriate to the age of the rocks concerned. For example, the ^{14}C method can be used for rocks up to about 50,000 years in age, which is useful in the Pleistocene.

Supposing, then, that today's direction of magnetic polarity is called "normal" (usually graphed in black) and the opposite is called "reversed" (depicted white), one can draw two parallel lines, marked with black and white areas, that indicate the (dated) geomagnetic reversals (Figure 12.1). The most recent geomagnetic epoch is called the Brunhes Normal Epoch, while the one preceding it is called the Matuyama Reversed Epoch. Further epochs are named in honor of scientists who played important roles in the history of geomagnetism, such as Gauss and Gilbert. The "fixed points" provided by geomagnetic polarity reversals have proved invaluable in chronostratigraphy, especially for the Pleistocene, but they become increasingly insecure for older rocks.

Needless to say, the geomagnetic time scale would get into a muddle if any reversals are missed (like bar codes if any bars are missing). Thus, the recognition of the brief so-called Jaramillo normal polarity event (named after a small river in New Mexico) within the reversed Matuyama Epoch was an important issue in the establishment of a reliable geomagnetic time scale in the lead-up to the plate tectonics revolution (Glen 1982).

Besides looking for geomagnetic reversals in exposed lava sheets, it was also desirable to try to find independent evidence for them in submarine

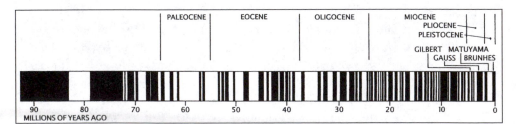

Figure 12.1: Part of the geomagnetic time scale (Hoffman 1988, 55). Artwork © Ian Walpole. Reproduced by courtesy of I. Walpole. (Only the right-hand part of his diagram is reproduced here.)

cores. Efforts were made in this direction in the 1950s by Maurice Ewing (1906–1974), director of the Lamont Geological Observatory, New York (now the Lamont–Doherty Earth Observatory), and Manik Talwani (then at Lamont but subsequently of Rice University, Houston), examining submarine cores obtained by Lamont's research vessel *Vema*. They took some cores to the Carnegie Institution, Washington, and asked researchers there to examine their magnetic reversals, but the cores were soft and the results ambiguous, and thus the matter was not pursued at that stage.

By the following decade, however, techniques had improved, and the Lamont Laboratory had an enormous "library" of drill cores from the world's ocean floors. These were examined by the Lamont rock magnetism specialist, Neil Opdyke, assisted by a graduate student John Foster and another student, Billy Glass. They were aided by James Hays, also at Lamont, an expert on the biozonation of small siliceous surface-dwelling organisms called radiolarians, found in the sediments under the Antarctic oceans, which provided paleonto-logical "control" of the sediment cores and a cross-check on the remarkable geomagnetic-reversal correlations demonstrated between terrestrial lava flows and deep-sea sediments (Opdyke et al. 1966).

Thus, the "geomagnetic polarity time scale" (GPTS) was established and linked to climatic changes, recognizable by the cold-water fossils species (radiolarians) in the cores—a faunal zonation for which was proposed (Hays and Opdyke 1967). The identification of a specific paleomagnetic reversal signal in a core made it possible to date a particular position in the core, and one could make additional time correlations for the Pleistocene to locali-ties elsewhere on the Earth. Thus, in the years that followed, it became pos-sible to have a "date-tagged" history of Pleistocene climate. Hays and Opdyke (1967) claimed an extension of their procedure back to about 5 million years. Moreover, they suggested that there had been pre-Pleistocene glaciation, and there was some (inconclusive) evidence that faunal changes were associated with geomagnetic reversals.

The date for the beginning of the Pleistocene (and other matters) was set-tled with the aid of geomagnetic investigations. The Eighteenth International Geological Congress in London (1948) had recommended that the beginning of the Pleistocene be defined by the appearance of certain cold-water species in southern Italy (Eighteenth International Geological Congress 1950, 6; King and Oakley 1950, 213–14). This followed the suggestion of Maurice Gignoux (1910) that the base of the Quaternary (or base of the Pleistocene, a term he did not use) should be defined by a site in Calabria, where the cold-water fos-sils (especially *Cyprina islandica*) first appeared. But how could someone else-where in the world make a correlation with a "marker" defined like that? The problem was solved when William Berggren of the Woods Hole Oceanographic Institution, Cape Cod, and James Hays showed that this Italian event coin-cided (near enough!) with the so-called Olduvai Normal Event (a short epi-sode of about 100,000 years within the Matuyama Reversed Epoch) at about 1.8 million years ago (Hays and Berggren 1971), which could be recognized globally. Thus, geomagnetic reversals provided a tool for recognizing the base of the Pleistocene.

There was another element in this process of providing a time scale based on the study of submarine cores, referred to briefly in Chapter 9. Climate changes, by definition, alter global temperatures. The proportions of the isotopes of oxygen and hydrogen in water, or the proportions of the different "kinds" of water of different isotopic compositions, vary according to the temperature. This is because the relative volatilities of the different forms of water will differ at different temperatures, and hence the atmosphere will contain more or less $H_2{}^{16}O$ and $H_2{}^{18}O$ at different average global temperatures, the two relevant oxygen isotopes being the predominant ^{16}O and the subordinate ^{18}O. Thus, the proportion of $H_2{}^{16}O$ and $H_2{}^{18}O$ differs in seawater according to the prevailing temperature. As a result, organisms forming calcium carbonate shells will have different proportions of ^{16}O and ^{18}O in their shells according to water temperature. Examination of the oxygen isotope compositions of the calcite of fossils from sediment cores taken from the ocean floors will thus indicate variations in climate and will do so more precisely than will the occurrence of particular fossil types.

It was Harold Urey (1947) who initially suggested that the uptake of ^{18}O and ^{16}O by calcite-shelled organisms would be dependent on the ambient temperature of the seawater, so in principle one might gauge former temperatures by examining the isotopic composition of fossils' calcite. Initially, the different uptake rate was thought to be the main factor involved in producing calcite of different isotopic compositions. But later it was recognized that the actual isotopic composition of water also changes with temperature because of the differential volatilities of $H_2{}^{16}O$ and $H_2{}^{18}O$, as discussed previously. Fortunately, the two temperature effects influence the isotopic composition in the same "direction" so that measurement of the isotopic deviation (δ) provides a viable method for estimating changes in water temperatures.

The isotopic deviation (δ) of a sample of an oxygen-containing substance such as calcite (calcium carbonate) from some "standard" value is defined by the following equation (Emiliani 1955, 340):

$$\delta^{18}O = 1,000\{[(^{18}O/^{16}O)_{sample} - (^{18}O/^{16}O)_{standard}]/(^{18}O/^{16}O)_{standard}\}$$

or

$$\delta^{18}O = 1,000\{(^{18}O/^{16}O)_{sample}/(^{18}O/^{16}O)_{standard} - 1\}$$

The $(^{18}O/^{16}O)$ standard is taken to be that possessed by a particular sample of seawater, kept at a fixed temperature in Vienna, which is thus termed the Vienna-Standard Mean Ocean Water. Values of $\delta^{18}O$ may be determined along the lengths of sediment cores, using a mass spectrometer to measure the proportions of the different isotopes present for each sample, and the results may be displayed graphically, revealing the climate changes during the time interval represented by the core. Since the values are proxies for temperatures, one can recognize different warm and cold "stages," working back from the present: 1 (warm), 2 (cold), 3 (warm), and so on. Following up the work in the Pacific of the Swedish investigator Gustaf Arrhenius (1952), Emiliani (1955) plotted temperature variations for up to 15 such stages, and the number has subsequently been increased. Note that Arrhenius's stages were not based on temperature comparisons per se. He analyzed the calcium carbonate content of deep-sea cores from the eastern Pacific, which showed troughs and peaks when graphed. He numbered these from 1 to 18, starting with 1 at the present. The peaks (2, 4, 6, . . .), he thought, represented glacial conditions, while the troughs (1, 3, 5, . . .) represented interglacials.

Emiliani (1966) subsequently proposed a series of paleontologically defined subdivisions (based on foraminifera) to be fitted alongside his graph of $\delta^{18}O$ plotted against time. These zones were necessary, for without the evidence of the evolving foraminifera, one would not know where one was (in time and for the purposes of correlation) on the basis of a cyclically varying plot of $\delta^{18}O$ versus time, ascertained for some sediment core. The 1966 paper proposed 17 stages for sediment cores from the Caribbean, dating back 425,000 years.

Likewise, it was found by Wallace Broecker and Jan Van Donk of the Lamont Laboratory that the measured values of $\delta^{18}O$ varied down the length of another core extracted from the seafloor in the Caribbean region (Broecker and Van Donk 1970), as shown in Figure 12.2.

Such patterns evidently indicated cyclic changes in the temperature of seawater. The evidence was thought to be in accord with Milanković's idea of climatic oscillation during the Pleistocene, though, with a periodicity of about 100,000 years, there was not an exact fit with what Milanković's theory required, and the fit wasn't exactly right for the older Ice Age terminations. So more and better data were sought.

To tackle the problem, three laboratories (Lamont, Brown University, and Oregon State University) pooled their expertise and resources, and researchers from several countries in Europe also became involved. The biostratigraphy needed to be meshed with the oxygen isotope cycles and the geomagnetic time scale, which provided a "time tag" for the whole (specifically at the Brunhes/Matuyama reversal). But initially there was difficulty in finding a core that extended down as far as that reversal. A suitable one was eventually found in the Lamont "core archive," collected in 1971 from the western Pacific: the Ontong–Java Plateau (Core V28–238). Hays found that it contained the desired reversal, and core samples were sent to Nicholas Shackleton at Cambridge (United Kingdom) for isotope analysis. This allowed subdivision of the Brunhes Polarity Epoch into 19 stages based on the isotopic curve (not on paleontological criteria, as for the lower parts of the stratigraphic column). The results were published by Shackleton and Opdyke (1973) (Figure 12.3). Imbrie and Imbrie (1979, 165) called this the "Rosetta stone" of the problem. The top of the core was datable by standard radiocarbon dating techniques, and a secure marker point near the bottom was provided by the Brunhes/Matuyama reversal. So V28–238 provided a means for the calibration of the time intervals (by interpolation) on the oxygen isotope curve. The length of the stages was found to be greater than Emiliani (1966) had originally proposed. Shackleton and Opdyke envisaged the stages as having global validity.

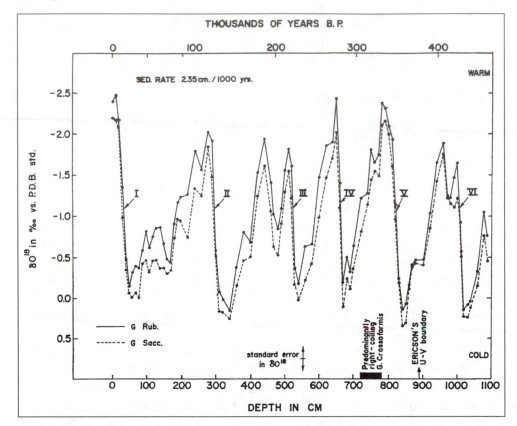

Figure 12.2: Variations of oxygen isotope composition with depth for two species of foraminifera for a Caribbean core (V12–122) (Broecker and Van Donk 1970, 175). The roman numerals indicate the times when there were sudden decreases in the value of $\delta^{18}O$, indicating the onset of a glacial episode. V stood for *Vema*, 12 represented its twelfth voyage, and 122 was the core number of that voyage. © 1970. Reproduced by permission of the American Geophysical Union.

There followed further efforts to link the subsidiary peaks of the isotope variation graph ($\delta^{18}O$ versus time) to the predictions of Milanković theory. Various difficulties were encountered (see Imbrie and Imbrie 1979, chap. 15), but by using several overlapping cores, it was found possible to produce a climate curve extending back some 450,000 years. And when this was subjected to spectrum analysis (see p. 216), cycles were found in conformity with the Milanković predictions, with periods of 100,000, 43,000, 24,000, and 19,000 years (Hays, Imbrie, and Shackleton 1976). In the words of Imbrie and Imbrie (1979), the "mystery of the ice ages" had seemingly been solved.

In more recent work, when there has been discrepancy between the age of a geomagnetic reversal as determined by radiometric dating and its age as determined by the mathematically calculated Milanković cycles, used in conjunction with the oxygen isotope curves and the paleontological evidence from sediment cores, primacy has been "awarded" to Milanković and the biostratigraphic and

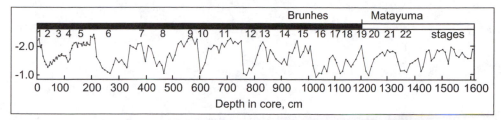

Figure 12.3: Comparison of oxygen isotope ratios and magnetic polarities for Lamont Core V28–238 (Shackleton and Opdyke 1973, 48, fig. 9). © 1973. Reproduced by permission of Elsevier. Redrawn by Ricochet Productions.

oxygen isotope evidence (e.g., Shackleton, Berger, and Peltier 1990; see also the review by Kent 1999). That is, the astronomical calculations are now used to "tune" (= calibrate) the sedimentary record and even the paleomagnetic record for relatively young rocks. And studies are being extended well before the Pleistocene, where attention was initially focused. This has been made possible by technological advances in drilling/coring techniques, with multiple holes being drilled per site. The geological time scale is now refined as far as the bottom of the Miocene, with the new boundary there being placed at a somewhat younger date than hitherto (Gradstein, Ogg, Smith, Blecker, and Lourens 2004).

It is noteworthy that astronomical calculations have "forced" some reappraisal of dates determined radiometrically, with the standards used in such datings being revised somewhat. Thus, the astronomers may be beginning to hold the whip hand over geologists and even geochemists and geophysicists! It is also to be noted from Gradstein, Ogg, Smith, Blecker, and Lourens (2004) that there has been a move to place the bottom of the Quaternary at the bottom of the highest stage in the *Pliocene,* not at the Pliocene/Pleistocene boundary. Thus, the Quaternary is in danger of being superseded!

There are also developments in studies of sedimentary cycles in ancient rocks that may have had astronomical causes. Linda Hinnov (2004) of The Johns Hopkins University has described "astronomical signals" in the sedimentary record for the Cretaceous and the Triassic and even for the Cambrian. Indeed, in a copy of the PowerPoint presentation of this paper, which she kindly made available to me, she referred to the possibility of such signals even being recognizable in the Precambrian, having reference to evidence for glacial and interglacial deposits in wadis in Oman. It remains to be seen how far back in time the astronomical signal can be used to "tune" the geological time scale.

But such recent issues are matters for geologists, not historians, to sort out. It is interesting, though, to see how current thinking is coming to be influenced ever more by the long-ignored Milutin Milanković. And the work described here has put the nail in the coffin of the theory of four ice ages of the early investigators Penck and Brückner, discussed in Chapter 9.

13

GEOCHEMICAL CYCLING

We come now to one of the most interesting aspects of geological cycling—in geochemistry. That is, we consider what can be learned about the way the Earth works as a system as different elements move from one region to another (from "sources" to "sinks"). The various elements may be found in and move between the atmosphere, the oceans, freshwaters (liquid or solid), sediments, and metamorphic or igneous rocks. They can be deep within the Earth in the mantle *and* in living organisms or their remains.

The idea of the hydrological cycle occurs in Aristotle. The double cycle of oxygen and carbon, involving respiration and photosynthesis, is well known. But when we consider the movements of all the different kinds of elements, in the various places where they may be located in the geocosm, we are dealing with a vast number of possible chemical and physical changes, which need to be understood and to some extent represented in idealized form (by "modeling") in order to be comprehended. Such problems were not treated closely until the twentieth century.

The study of the geocycles of elements other than carbon and oxygen (and to some extent nitrogen and phosphorus by agriculturalists) began with sodium. The first thought in this direction had to do with a method for estimating the age of the Earth. The astronomer Edmond Halley (1656–1742) (1715) suggested a procedure that assumed that salt gradually accumulates in lakes or oceans over time. The waters started off fresh but gradually became saline by the transport of sediment and dissolved salt from land to sea. He proposed that the salinity of lakes or inland seas be determined every hundred years, and hence the age of such bodies of water could be estimated. If the scheme were applied to the oceans, one might get an approximate idea of the Earth's age. Obviously, the scheme assumed that there was no "back-cycling" from oceans to land. It was not taken up in Halley's day.

However, the Liverpool architect, engineer, and amateur geologist Mellard Reade (1832–1909) published work that was in principle similar to what Halley envisaged. Reade (1877) estimated that it would have taken 200 million years for the oceans to have acquired their present salinity. Analogous work was undertaken by the geology professor at Trinity College, Dublin, John Joly (1857–1933), who suggested, on the basis of the estimated rate of transfer of sodium to the oceans, that the Earth was about 90 million to 100 million years old (Joly 1899). He made the dubious "uniformitarian" assumption that the rate of sodium increase in the oceans was approximately constant. Nevertheless, his figure for the Earth's age was quite pleasing to geologists. They were being pressed at that time by physicists, notably Lord Kelvin, who contended, on the basis of astronomical arguments and the quantity of heat residing in the Earth, that the planet might only be about 20 million years old. Joly later (1930, 152) contemplated 240 million to 350 million years, but by that time the salinity method for estimating the Earth's age was obsolete, being superseded by radioactivity determinations. Joly also failed to take account of the recycling of sodium from the oceans into sediments by the evaporation of "inland" seas or by salt spray being blown inland. And his estimates were wholly out of line with those indicated by radiometric work, which, quite early in the twentieth century, suggested that the Earth was a billion years old or more (Lewis 2000).

During the nineteenth century, German petrologists such as Harry Rosenbusch (1836–1914) developed what came to be called "classical magmatism" (Gregor 1988, 1992). This tended to move away from Huttonian cycles, which entailed the formation of fluid magma by the recycling of buried and heated sediments. For the "classicals," the loss of igneous rock by weathering and deposition of sediment was made good by eruptions of magma from the Earth's still partly fluid interior, which was "juvenile"—that is, "new" magma that had never seen the light of day previously. However, they agreed that sediments could be converted to metamorphic rocks by heat and pressure, and so too could igneous rocks.

There were also the "migmatists"—mostly in Scandinavia—who thought that some metamorphic rocks (e.g., gneisses) were produced by a kind of hot solution (called "migma" or "ichor") that could soak into sediments, causing metamorphism. If the process proceeded sufficiently, there could be complete conversion into granite, which, when produced thus, could be thought of as a kind of hypermetamorphic rock (my terminology). Such ideas, though the source of controversy between migmatists and magmatists (Read 1957), had in common the idea that magma, migma, or ichor were not recycled in some "Huttonian" manner.

Early in the twentieth century, however, geochemical evidence emerged that a noncyclic cause of the oceans' salinity was inadmissible. Frank Clarke (1847–1931), chief chemist at the U.S. Geological Survey, issued a series of bulletins that compiled the results of numerous chemical analyses of igneous rocks (Clarke 1908 and four subsequent editions). With this information, he was able to calculate the approximate amount of sediment corresponding to the transfer of the quantity of sodium thought to be present in the oceans. (The estimation assumed that weathering of igneous rocks "split" them into sediments and sodium compounds.) On this basis, it seemed that the seas' sodium corresponded to less than a thousand meters of sediment (if distributed uniformly over the Earth's present land surface). But this amount of sediment and the oceans' salt could have been formed since the Cretaceous. Something was wrong somewhere. It would seem, then, that the sodium in the oceans had to be recycled into the rocks, providing material for subsequent weathering. In the light of plate tectonic theory, it is evident that such cycling of sodium is possible, as sodium-bearing sediments are drawn into troughs during subduction. But that route was not much accepted before the 1960s, and the "sodium problem" was brushed aside for some years.

It was, however, attended to by the Dutch geologist Willem Nieuwenkamp (1903–1979) of the University of Utrecht, who raised the question at the Eighteenth International Geological Congress, held in London in 1948 (Nieuwenkamp 1948). Repeating Clarke's calculation, which had also been performed by his teacher, the geochemist Victor Goldschmidt (1853–1933), Nieuwenkamp argued the following:

> The total mass of sodium in the oceans is
>   approximately 1.5×10^{22} g.
> 100 g of average igneous rock contains
>   2.8 g of sodium.
> 100 g of average sedimentary rock contains
>   1.0 g of sodium.

Therefore: 100 g of sediment derived from igneous rock
                      delivers 1.8 g of sodium to the oceans.

Therefore: 1.8/100 g of sodium is delivered to the oceans
                      when 1.0 g of sediment is formed.

Therefore: the weight of sedimentary rock needed to
                      produce the mass of sodium in the oceans is
                      $(1.5 \times 10^{22} \times 100)/1.8$
                      $= 85 \times 10^{22}$g.

Such a quantity would be equivalent to about 2,000 meters of sediments spread over the Earth's land surface. But the quantity of sediment was evidently much greater than that. Therefore, there should be more sodium in the oceans than that which is actually found. So the sodium must somehow be removed from the oceans. "One is tempted to think of circulation," wrote Nieuwenkamp. He was returning to Huttonian theory and away from "classical magmatism."

Considering the geochemical relations of other elements also, Nieuwenkamp explored the following generalization: every rock or mineral has, at some stage in its process of formation or cyclic history, existed in the form of sediments. Rocks, whether sedimentary, igneous, or metamorphic, are at some stage all

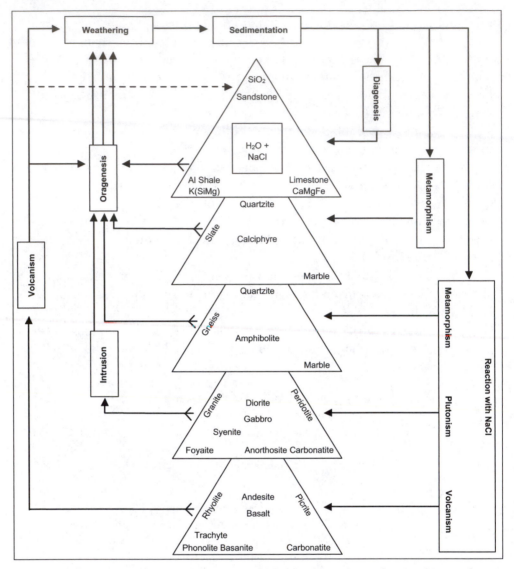

Figure 13.1: Nieuwenkamp's "persedimentary" model of the continuous rock series (Nieuwenkamp 1968, 366, fig. 1). Uppermost triangle: sediments; second triangle: metamorphosed sediments; third triangle: metamorphic rocks; fourth triangle: coarse-grained igneous (plutonic) rocks; bottom triangle: finer-grained igneous rocks. The entities on the right side of each triangle are basic (low in silica); those at the left and top are more acidic (silica rich). Reproduced from *Geologische Rundschau* by permission of Springer-Verlag. Redrawn by Ricochet Productions.

formed via sediments. Hence he called his view the "*per*sedimentary" model and represented it diagrammatically as shown in Figure 13.1.

These ideas were promulgated at about the time that the plate tectonics revolution was in full swing, and according to that theory, one would indeed expect sediments to be carried down into the Earth's nether regions (mantle) and be recycled. But Nieuwenkamp arrived at such ideas from a *geochemical* standpoint, not from ideas of seafloor spreading, seismological evidence for subduction, or whatever. Although enunciated chiefly through three papers

(1948, 1956, 1968) rather than some classic book, his ideas influenced several prominent geologists, such as Arthur Holmes, Eugene Wegmann, and Thomas Barth. And his ideas and reputation were actively applied and promoted by Bryan Gregor (1966, 1988, 1992) of Wright State University, Ohio, who worked for a time under Nieuwenkamp in the Netherlands.

It is now thought that the reactions occurring on Earth in its early stages were significantly different from those occurring today. The early Earth (formed about 4.5 billion years ago) would have been much hotter than at present and presumably without oceans. After the solidification of a crust, an atmosphere could have formed by "outgassing" from volcanoes. It would have created a "reducing" environment, with substances such as water vapor, carbon monoxide, carbon dioxide, methane, ammonia, nitrogen, hydrogen chloride, and sulfur compounds. With cooling, oceans could form, but the conditions would have been very different from today because of the absence of life. In the later parts of the Archean (3.6 to 2.5 billion years ago), we find the first surviving forms of life (bacteria about 3.5 billion years ago). During the Proterozoic (2.5 to 0.54 billion years ago), cyanobacteria became common and liberated oxygen to the atmosphere, producing (at about 2.2 billion years ago) oxidizing as opposed to reducing conditions. At that time, we find iron oxides in fossil soils and in great iron ore deposits such as those of Western Australia. In the last part of the Proterozoic (the Ediacaran Period: 0.63 to 0.54 billion years ago), there were macroscopic, soft-bodied organisms that have survived as fossils in some parts of the world, such as Ediacara in South Australia, where they were first recorded. The previously mentioned changes were "one-way" or noncyclic, though the photosynthesis involved chemical cycles. The Earth was also cooling, for although radioactive minerals liberate heat, their supply slowly declines.

The foregoing subdivisions of the stratigraphic column together make up the Precambrian. The Earth's magnetic field was established during this early time, perhaps when the inner core formed and convection cells were established in the core. Plate tectonic activity is thought to have been taking place during the Precambrian. With the Cambrian that followed (at 0.54 billion years ago), we have the first "shelly" fossils and the stratigraphic column (Phanerozoic = Paleozoic + Mesozoic + Cenozoic) up to the present, to which time span geologists have devoted most of their attention.

Let us now attend to what geochemistry may do to help us understand Earth history. The term "geochemistry" was defined by Goldschmidt (1958, 1) as "studies [of] the distribution and amounts of the chemical elements in minerals, ores, rocks, soils, waters, and the atmosphere, and the circulation of the elements in nature on the basis of the properties of their atoms and ions." The term itself was first introduced in 1838 by the Basel professor Christian Friedrich Schönbein (1799–1868) (1840), an electrochemist and discoverer of ozone (O_3). He thought that the Earth's inorganic materials were deposited according to knowable chemical laws, so that in principle the *chemist* might write the history of the globe. This idea appealed to Carl Gustav Bischoff (1792–1870), professor of chemistry at Bonn, who presciently regarded the

Earth as a vast chemical laboratory, with ongoing cycles of the elements. He collected data on the chemical compositions of gases, waters, minerals, and rocks but couldn't answer Schönbein's call. In Canada, the Geological Survey officer Sterry Hunt (1826–1892) (1875) endeavored to provide a chemical "just-so story" for the Earth's chemical history. But his efforts were premature and were derided by most of his contemporaries.

Schönbein's program was revived in Russia by the chemist, mineralogist, and crystallographer Vladimir Ivanovich Vernadsky (1863–1945), professor at Moscow and then at St. Petersburg. He was a student of Dmitri Mendeleev, of periodic table fame, and the soil scientist Vasili Dukuchaev, but his ideas came to him during World War I when he was gathering information on Russia's strategic raw materials. This led him to consider the size of the global repositories of different elements, and he realized the significance of living organisms (including humans) in geological processes. His thoughts turned in the direction of a global ecology, or the ecology of the biosphere (the term earlier proposed by Suess), and he coined the term "biogeochemistry." Vernadsky spent some of his early years in Paris, where he was influenced by the philosopher of evolution Henri Bergson. In consequence, some of Vernadsky's ideas were first published in French (1924, 1929) and are also now available in English (1997).

Vernadsky showed that life penetrated into all corners of the globe, from the upper atmosphere to the planet's deep interior, and that life was involved in one way or another in nearly all sedimentary rocks—coals, limestones, salt deposits, phosphates, ironstones, shales and mudstones, and so on—and ultimately in all geological changes. Moreover, the biosphere formed a system that was itself analogous to a living organism in that it was self-regulating. To study this gigantic system, one needed to examine quantitatively the various "sources" and "sinks" for the different elements, the ways in which matter was transferred between them, and the cycles involved in the exchanges. All this lay beyond Vernadsky, especially since much of his career was spent in war years or periods of political upheaval. But his ideas were revived in the 1960s and made headway in the United States partly through the efforts of an entomologist disciple, Alexander Petrunkevich, and Vernadsky's historian son, Georgii Vladimirovich (both of whom worked at Yale), and the Yale zoology professor George Evelyn Hutchinson.

Here, then, we turn to the development of "Vernadskyan" ideas in the United States by geochemists proper, notably Robert Minard Garrels (1916–1988). He started at Northwestern University (1942) and subsequently became chief of the Solid States Group of Geochemistry and Petrology at the U.S. Geological Survey (1952). From there he moved to Harvard (1955), back to Northwestern (1965), then to the Scripps Institute in California (1969), into the Pacific to Hawaii (1972), back again to Northwestern (1974), and finally to the University of South Florida (1980). He was always on the move—like the elements he studied.

By the second half of the twentieth century, with the advances in chemistry since the time of Sterry Hunt, it was possible to suggest chemical equations for reactions that might have been operating in the interconversions of compounds in Archean times. For example (Garrels and Mackenzie 1971, 294),

$$14\,H_2O + 6NaAlSi_3O_8 + 3FeCO_3 + 15MgSiO_3 + 3CaCl_2 =$$
$$\text{Albite} \qquad \text{Siderite} \quad \text{Enstatite}$$

$$22SiO_2 + 6NaCl + 3Mg_5Al_2Si_3O_{10}(OH)_2 + 3CaCO_3 + Fe_3Si_2O_3(OH)_4$$
$$\text{Silica} \qquad \text{Chlorite} \qquad\qquad \text{Calcite} \quad \text{Greenalite}$$

$$5MgSiO_3 + CaAl_2Si_2O_3 + CO_2 = Mg_5Al_2Si_3O_{10}(OH)_8 + CaCO_3 + 4SiO_2$$
$$\text{Enstatite} \quad \text{Anorthite} \qquad\qquad \text{Chlorite} \qquad\qquad \text{Calcite} \quad \text{Silica}$$

$$Fe_3Si_2O_3(OH)_4 + 3CO_2 = 3FeCO_3 + 2SiO_2 + 2H_2O$$
$$\text{Greenalite} \qquad\qquad \text{Siderite} \quad \text{Silica}$$

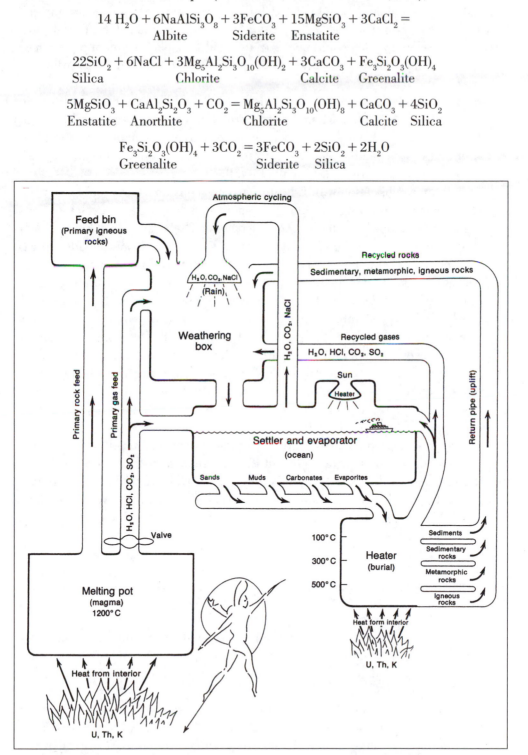

Figure 13.2: The "crust–ocean factory" (Garrels and Mackenzie 1971, 330). © Cynthia Garrels.

In such reactions, the iron is in its lower "oxidation state"—that is, in its "reduced," "ferrous," or "divalent" condition.

But, as mentioned, at about 2.2 billion years ago, iron made its appearance in its oxidized or "ferric" condition (trivalent), with deposits of red-brown hematite (Fe_2O_3). This could be associated with the establishment of an atmosphere containing oxygen due to the photosynthetic activities of primitive organisms.

In the same book, Garrels provided a striking diagram representing the cyclic processes that occur in, on, and around the Earth, as shown in Figure 13.2. The energy driving such processes in the "crust–ocean factory" would be supplied by radioactive elements within the Earth, plus the energy supplied by the Sun, and captured during photosynthesis.

In a later publication with Edward Perry (Garrels and Perry 1974), Garrels provided an interesting diagram that again depicted various cycles operating in and on the Earth and in the atmosphere. Their "cartoon" (Figure 13.3) showed the reservoirs for different materials—oxygen, carbon dioxide, silica, gypsum, organic matter, and so on—and their supposed interconnections, the volume of each reservoir in the diagram being proportional to the estimated number of molecules in each reservoir. Numbers on the connecting links indicated the supposed rates of movement (fluxes) of matter from one reservoir to another. For example, the quantity of carbon in the atmosphere's carbon dioxide reservoir is minute compared with the other reservoirs, but CO_2 is cycling very rapidly. By contrast, the silica reservoir is huge, but the exchange of silicon with other reservoirs proceeds slowly. Some of the reactions were involved in weathering, diagenesis (changes in sediments after deposition), or metamorphism.

The system is presumably broadly stable over time, but it was interesting to know whether it was constant through geological time, at least during the Phanerozoic. To investigate this question, Garrels went on, with Abraham Lerman of Northwestern, to investigate the carbon and sulfur cycles and their possible *coupling*. The proposed basic equation (hereafter Equation 1) for the exchanges was as follows:

$$4FeS_2 + 8CaCO_3 + 7MgCO_3 + 7\ SiO_2 + 15\ H_2O = 15'CH_2O' + 8CaSO_4 + 2Fe_2O_3 + 7MgSiO_3$$

Pyrite   Limestone/Dolomite   Silica   Organic   Gypsum   Hematite Enstatite
                                                  compounds

It will be remarked that iron and sulfur are oxidized as the equation shifts from left to right, and the limestone/dolomite reservoir gives way to gypsum. The converse holds if and when the equation shifts from right to left. Sulfur, we note, may be bound up in either pyrite or gypsum, and carbon can occur in limestones or in organic compounds (loosely represented by CH_2O). Can such interchanges be occurring systematically through geological time? It is striking that we have large limestone deposits in the Carboniferous, while gypsum and other evaporites are prominent in the Permian and Triassic. Is this just fortuitous?

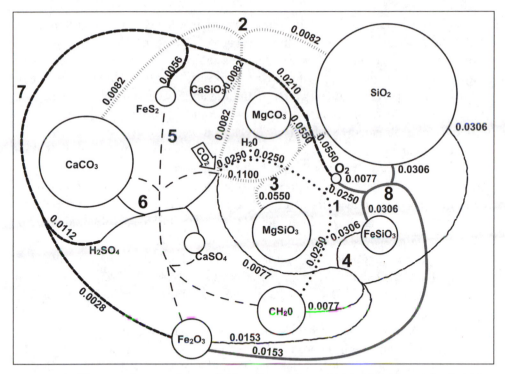

Figure 13.3: The "steady-state flow net" (Garrels and Perry 1974, 313). Copyright © 1974 John Wiley. Reproduced by permission of John Wiley & Sons.

The single-digit numbers in the diagram refer to the following reactions:

1 $CH_2O + O_2 = H_2O + CO_2$

2 $CaCO_3 + SiO_2 = CaSiO_3 + CO_2$

3 $MgCO_3 + SiO_2 = MgSiO_3 + CO_2$

4 $2Fe_2O_3 + 4SiO_2\ CH_2O = 4FeSiO_3 + H_2O + CO_2$

5 $2Fe_2O_3 + 15H_2O\ 8CaSO_4 = 8CaCO_3 + 4FeS_2 + 7CO_2 + 15H_2O$

6 $CaCO_3 + H_2SO_4 = CaSO_4 + H_2O + CO_2$

7 $4FeS_2 + 8H_2O + 15O_2 = 2Fe_2O_3 + 8H_2SO_4$

8 $4FeSiO_3 + O_2 = 2Fe_2O_3 + 4SiO_2$

To investigate such matters, Garrels and Lerman drew on data on the isotopic compositions of carbon and sulfur from the different reservoirs for rocks of different ages in the Phanerozoic, that is, on measurements of $\delta^{13}C$ and $\delta^{34}S$ for rock samples from the different periods of the Cambrian and up.[1] During the interchanges between the reservoirs, there is fractionation of the two isotopes of each element. For example, the fractionation of sulfur between pyrite and gypsum is temperature dependent. In addition, in organic processes, it is easier to break ^{32}S–O bonds than ^{34}S–O bonds. Thus, the *lighter* sulfate ions are more easily converted to sulfide ions. So sedimentary sulfides are preferentially less dense and therefore have negative values of $\delta^{34}S$. But the proportion of ^{34}S increases when sulfide ions (S^{2-}) are oxidized to sulfate ions (SO_4^{2-}), and $\delta^{34}S$ is then positive.

Garrels and Lerman (1981) developed a model for the carbon and sulfur cycles and the geological history of sulfur and carbon reservoirs considering just five reservoirs: the ocean/atmosphere (containing sulfates and carbonates in solution), gypsum deposits (oxidized sulfur), pyrite deposits (reduced sulfur), limestone or carbonate deposits (oxidized carbon), and organic carbon compounds (reduced carbon). They could estimate the values of $\delta^{34}S$ and $\delta^{13}C$ for the reservoirs as they stand today (assumed to refer to prehuman but geologically Recent time), and the approximate numbers of molecules in each reservoir. Further, they had an idea of the "fluxes" (rates of transfer) of materials from one reservoir to another, with the conversions between the solid-state reservoirs all passing through the ocean/atmosphere reservoirs. And they knew the *changes* in the values of $\delta^{34}S$ and $\delta^{13}C$ as material passes from each of the four "solid" reservoirs into or out of the ocean/atmosphere reservoir.

They further hypothesized that the carbon and sulfur cycles were *coupled,* as would be the case if the changes were all summarized by Equation 1. And any changes in the fluxes would translate into changes in the values of $\delta^{34}S$ and $\delta^{13}C$ for the materials in the reservoirs.

Then average values of $\delta^{34}S$ and $\delta^{13}C$ for materials in the reservoirs were determined (empirically) backward in time for the known reservoirs for rocks from different periods of the Phanerozoic, and a curve was plotted back in time for $\delta^{34}S$. From this, it was then possible to *calculate* a curve for average values of $\delta^{13}C$ and compare the results with the empirically determined values found in the stratigraphic record. The results were necessarily only approximate and relied on assumptions such as that the size of the ocean/atmosphere reservoir remained unchanged through the Phanerozoic. Nevertheless, the empirical findings for the $\delta^{13}C$ values were in satisfactory agreement with the values obtained by calculation from the $\delta^{34}S$ curve. Thus, the geochemical coupling of sulfur and carbon through time seemed to be supported.

However, the argument was revised in a subsequent paper by Garrels and Lerman (1984), following an objection to the earlier reasoning put to them orally by Robert Berner of Yale in 1981. According to Peter Westbroek (1992, 105), quoting Garrels's recollections, Berner tried to hypothesize various conditions in the Cambrian and ran the Garrels/Lerman algorithm (mathematical model) forward in time from the Cambrian through to the present. Then he adjusted the variables for the Cambrian until, when the model was run, the correct modern values were found. He was able to do this, *but* his values for the Cambrian differed from those obtained by Garrels and Lerman, calculating backward from the present to the Cambrian. Something was wrong.

Garrels pondered the problem and found a solution in a dream! He imagined a film played backward so that, for example, a swimmer might dive *out* of a pond. Traveling back into the past by calculation would be analogous. So Garrels and Lerman reversed all the signs in their equations and reran their computer simulations. This time they got results for the Cambrian that agreed with Berner's hypothesized assumptions. Other investigators, such as Manfred Schidlowski and Christian Junge (1981) of the Max Planck Institute of Chemistry at Mainz, were involved in similar work, and Garrels and Lerman (1984) were able to show that the several investigators were obtaining comparable results for the magnitudes of the gypsum reservoirs in time.

One of Garrels's assumptions during the work outlined above was that the atmosphere's composition was approximately constant during the Phanerozoic—otherwise, animals would have been unlikely to survive. But the large reservoirs of solid carbon and sulfur compounds were slowly changing back and forth. When pyrite is converted to gypsum, oxygen has to be sourced from the oceans/atmosphere. But this is replenished by oxygen, in the last analysis sourced from limestones, so the oceans/atmosphere can retain an approximately constant composition. Thus, metaphorically, one could think of the four reservoirs "breathing" in a coordinated fashion.

The main fact to emerge from the foregoing was perhaps the interaction and correspondence of the carbon and sulfur cycles, which can be demonstrated by plotting the values of $\delta^{34}S$ and $\delta^{13}C$ against time and comparing the two graphs. For our purposes, this can be done conveniently by comparing two graphs from a paper by William Holser et al. (1988) (Figure 13.4), from which it can be seen that there is indeed coupling between the carbon and sulfur cycles. The curves are consonant with the requirements of Equation 1: a leftward trend in the left-hand curve roughly matches a rightward trend in the right-hand curve, with the two trends corresponding to an overall shift from left to right in Equation 1. The curves mark the ponderous coordinated "breathing" of the sediment reservoirs! And the sulfur curve records the well-known occurrence of extensive evaporite deposits, such as gypsum, found in the Permian and Trias. But note that the grand changes are the summary product of a large number of faster changes that are going on in respiration and photosynthesis, combustion, the hydrological cycle, the formation of sediments, and their weathering away.

There remains the question of what may have been "driving" these chemical changes in the geological past. Westbroek (1992, 97) has recorded that Garrels "began to liken the earth system to a grandfather clock, in which the rotation of one wheel determines the turning of all the others." What, then, were weights that drove the "clock" and the "pendulum" that somehow regulated it?

A likely candidate is the slow cycles associated with plate tectonic theory, perhaps those linked to the coalescence and dispersal of continents, and changes of sea level driven by tectonic agencies. These factors were considered by Fred Mackenzie, then at the University of Hawai'i at Manoa, and his doctoral graduate John Pigott, who became a basin analyst with Amoco International and is now at the University of Oklahoma (Mackenzie and Pigott 1981). This paper was utilized by Holser et al. (1988), and Mackenzie and Pigott's paper drew on Pigott's Ph.D. dissertation (1981).

Most geologists agree that in the Cambrian there were several separate continents that gradually moved together to form the single continent, Pangea, by the Permian (at which time there was a substantial glacial episode) and

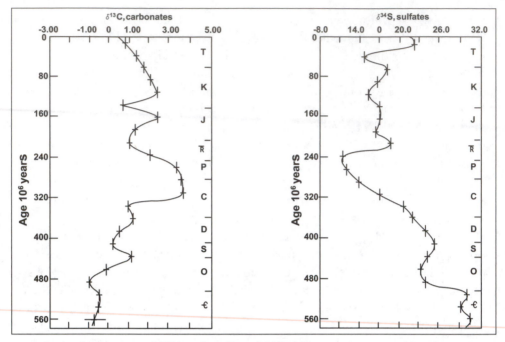

Figure 13.4: Comparison of δ¹³C carbonate and δ³⁴S sulfate isotope curves for Phanerozoic time (Holser et al. 1988, 139, 140). Copyright © John Wiley. Reprinted with permission of John Wiley & Sons. C = Cambrian, O = Ordovician, S = Silurian, D = Devonian, C = Carboniferous, P = Permian, Tr = Triassic, J = Jurassic, K = Cretaceous, T = Triassic. In the left-hand diagram, a trend to the left corresponds to the transfer of sulfur from pyrite to gypsum; in the right-hand diagram, a trend to the right represents the transfer of carbon from limestone to organic carbon. Redrawn by Ricochet Productions.

that Pangea began to break up during the Triassic, as envisaged by Wegener. Mackenzie and Pigott plotted the proportions of reduced carbon to total carbon and reduced sulfur to total sulfur, through Phanerozoic time, and found that they could be correlated. In his dissertation, Pigott (1981, fig. 15) had also correlated $\delta^{13}O_{oxidized}$ (carbonates) with $\delta^{13}O_{reduced}$ (shales), and these values could be correlated with the ratio of reduced carbon to oxidized carbon through the Phanerozoic. This in turn appeared to correlate with the rises and falls of sea level, given by the Vail curve (Figure 13.5). Thus, the grand driver for the geochemical changes might the movements of tectonic plates.

The correlation between geochemical changes and the sea-level curve given in Mackenzie and Pigott (1981) appears somewhat less persuasive (to me). Be that as it may, the authors linked the geochemical data with the Vail curve and suggested a scenario that might account for the relationship. Thus, sea level was thought to be high during plate convergence, when there were large submarine ridges. This led to the formation of widespread shallow seas around the continents, precipitation of limestone (from coral reefs in the shallow "epicontinental" seas), pyrite formation, and the uptake of gypsum and organic carbon (leftward shift of Equation 1). With fewer terrestrial organisms, there was a reduced supply of nutrients to the oceans but increased respiration

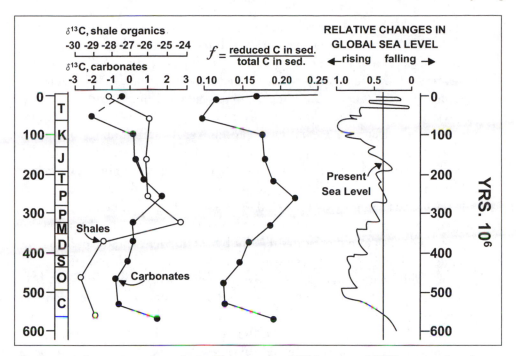

Figure 13.5: Comparison through the Phanerozoic of $\delta^{13}O_{oxidized}$ and $\delta^{13}O_{reduced}$, reduced carbon/total carbon, and sea levels (Pigott 1981, fig. 15). Redrawn by Ricochet Productions. Reproduced by permission of John Pigott.

globally. The consequent increase of atmospheric carbon dioxide led to further carbonate precipitation, and by "greenhouse principles" there would be elevated global temperatures. Under such high-temperature conditions, the oxygen content of estuarine waters and so on would be reduced and sulfides (pyrite) formed.

By contrast, when plates are dispersed and cratons elevated and oscillating in relation to sea level, there is decreased seafloor spreading and oceanic ridge volume. Overall, sea levels fall, reef carbonates are eroded, and there is more land available for plant growth. Sulfides are exposed and oxidized, and the sulfates thus formed are precipitated as "evaporites" (gypsum and so on), which accumulate in intracontinental basins. There is net consumption of carbon dioxide (global temperature falls), and organic carbon deposits increase. In sum, Equation 1 shifts to the right. Such conditions are associated with cyclothem formation.

There is another matter for consideration before we wrap up the geocycles in one large bundle! Westbroek has recorded that Garrels was puzzled by the stability of the whole system. Why didn't the "pendulum" swings get wildly out of hand? Was there "a haphazard interplay of physical and chemical forces"? Or was the whole system somehow organized and self-sustaining?

The answer toward which Garrels seems to have been moving and that Westbroek definitely advocates was that living organisms stabilize the whole. They survive in conditions that seem utterly unfavorable to life. Penguins survive on the ice in Antarctic winters. Fish live at extremely high pressures at the ocean floors. Bacteria occur deep within the Earth at high temperatures. Organisms are found in volcanic vents, in extremely acidic or alkaline conditions, or even on the control rods of nuclear power stations! Life is tenacious and almost ubiquitous on Earth (apart from its fiercely hot interior).

Living forms seemingly "behave" as if to "protect themselves." For example, when a volcanic island in the tropics sinks, its surrounding corals grow upward and maintain their optimum depth below sea level. And their reefs provide rich habitats for other organisms. Back in the Precambrian, carbonate solutions in rivers, produced by the dissolving of terrestrial limestones, passed into the oceans and provided material for the formation of carbonate platforms, formed by algal mats with (probably) photosynthetic bacteria. This material was carried down into the Earth's interior by subduction, and the carbon compounds eventually reemerge, in some cases even as carbonate magmas (see p. 174). Phosphatic materials are likewise concentrated by birds and bats, forming guano, which is eventually lithified and subsequently gets circulated by plate tectonic activity. If excessive carbon dioxide is liberated, this will partly be absorbed by dissolving in the oceans and by an increased uptake in photosynthesis. The atmosphere's oxygen does not, however, increase so much that materials become spontaneously inflammable: it is absorbed by the respiration of living organisms. Everything is cycling, quickly or slowly, and a self-sustaining system is maintained. Some authors, most famously James Lovelock (b. 1919) (1979, 1988), use the name *Gaia* (the Greek Earth goddess) to refer to the giant, complex, apparently self-sustaining system that we have on our planet, which can be regarded—by some literally and by others metaphorically—as a *quasi*-living organism. So the "pendulum" discussed in the present chapter does not swing out of control because of the involvement of living organisms stabilizing the whole. Of course, it is living organisms that facilitate exchanges between the element reservoirs, and Earth's heat loss is diminished by the blanket of its atmosphere generated by living organisms. The ozone layer shrouds us from dangerous radiation from space. It too is an indirect product of life.

So perhaps, modulated with the assistance of life, the tectonic and geochemical cycles are causally linked. And on top of them there are changes wrought by Milanković cycles.

NOTE

1. The $^{13}C/^{12}C$ standard is taken as that found in the calcite of a belemnite specimen known as the Pee Dee Belemnite, from the Cretaceous Pee Dee Formation, South Carolina. The standard specimen was formerly located at Chicago. It is now exhausted, but before then it was used to calibrate some secondary standards. (It has also been used for an $^{18}O/^{16}O$ standard.) The relevant sulfur isotopes are ^{34}S and ^{32}S. $^{34}S/^{32}S_{standard}$ is the value found in a troilite sample (FeS) from the Canyon Diablo Meteorite, Arizona.

14

THEMES AND VARIATIONS

We have now traced the general history of cyclic theories in geology from ancient to modern times. Cyclicity has appeared repeatedly, and one can hardly open any geological text today without seeing diagrams that display regular or irregular cycles. The idea of cyclicity is an example of what Gerald Holton (1978) has called *themata* (singular: *thema*). These are concepts or preconceptions (mostly unspoken) that guide scientists' imaginations and their work. Examples are atomicity (or its converse, continuity), simplicity (complexity), evolution (catastrophism), mechanism (vitalism), and chance (necessity). Sometimes they cannot be proven one way or another, and they may not in themselves be strictly scientific (e.g., the *themata* of divine creation or the infinity of space). Or they may be scientific principles that are now so well established (e.g., the laws of thermodynamics) that they have lost their initial hypothetical character.

Themata should not be conflated with the "paradigms" *(paradigmata)* of Thomas Kuhn (1962). In his famous work, Kuhn discussed groundbreaking pieces of scientific work that served as "exemplars" for others' investigations. A Kuhnian paradigm is *not* just a theory or an idea: it encompasses both a theory and the practical procedures involved in deploying that theory. But the word "paradigm" has become so loosely used that it often means little more than a concept that many people believe in, and it has found its way into political discourse, discussions in art history, or whatever.

Geoscience is today so permeated by the ideas of cyclicity that it doesn't have to be spelled out as something special, whereas for Hutton cyclicity required extended argument. For us, an explanation that invokes cycles is likely to be prima facie plausible. But cycles have not always been part of the geoscientist's armory, and, of course, they don't explain everything. In addition, there can be debate as to the correct application of a cyclic model and whether a cyclic process can perform the "work" required of it. (For example, are Milanković cycles adequate to account for important features of the stratigraphic column? Could Lyell's Great Year account adequately for glacial epochs?)

Among the important cyclic concepts we have encountered, I would specially mention Hutton's geostrophic cycle, the Wilson cycle in tectonics, geochemical cycling, sedimentary and geomorphological cycles, and Milanković cycles. One should also include the convection currents in the Earth's outer core and the much slower convection in the mantle. By bringing such matters to the fore in the present book, I have endeavored to present a historical account of some of the most interesting features of modern geology. Some issues of a cyclic nature have, however, been omitted for want of space (e.g., soil cycling, or the processes leading to the formation of layered igneous rocks, as seen, for example, in Greenland).

But there are many parts of geology where "cyclism" has little relevance. The evolution of living organisms is a kind of "one-way street." There may be repeated evolutionary "discoveries" of the same structure (e.g., cephalopod and vertebrate eyes). But the reemergence of old life forms, if similar conditions present themselves (e.g., Lyell on the iguanodon), has been rejected in "Dollo's law": evolution is irreversible. Evolution seems to proceed by a "random walk" or according to a pathway governed by natural selection, not by cycles. Volcanoes do not seem to erupt according to any law-like cyclic pattern. Earthquakes cannot be foretold satisfactorily, as they do not follow definite patterns in time, though their spatial distributions are known. Mineralogists and crystallographers don't make much use of cyclic notions.

In Chapter 13, I discussed the geochemical evidence for a kind of controlled cycle of changes in the stratigraphic column in the Phanerozoic, the motor for the whole being the movements of plates and the Wilson cycle, itself propelled by mantle convection. The swings in the changing "geocircumstances" are arguably modulated by living organisms, which "buffer" the changes. Possibly, plate tectonic activity, as we know it, is dependent on the existence of living organisms—or tectonic changes would be quantitatively and qualitatively different without life on Earth.

However, some authors (e.g., Frankel 1999) believe that the different periods of the Phanerozoic, marked by substantial changes in life forms (or death forms: fossils), have arisen because of impacts from extraterrestrial objects ("bolides"). This view is not universally accepted, though many geologists believe that the great extinctions at the end of the Permian and the Cretaceous were due to bolides. Astronomical impacts are, one might suppose, random events, and Frankel does not claim that a cyclic pattern is involved, though he maintains that most of the major boundaries in the stratigraphic column are marked by the abnormal presence of rare elements (especially iridium), derived from extraterrestrial objects, or by actual craters, especially one by the Yucatán Peninsula (Mexico), thought to mark the end of the Cretaceous.

However, a *periodicity* of 26 million years for episodes of high incidence of bolides has been proposed by, among others, David Raup and the late Jack Sepkoski of the University of Chicago (1984) and Marc Davis, Piet Hut, and

Richard Muller (1984) of the Lawrence Berkeley Laboratory, University of California, and Princeton, who hypothesized an (invisible) companion star to the Sun, dubbed "Nemesis" (after the Greek goddess who persecuted the rich, proud, and powerful). Supposedly every 26 million years, this "thing" came fairly close to Sun and passed through its large "cloud" of comets called the Oort cloud. When this occurred, some of the comets were deflected Earthward. The empirical evidence for the repeated extinctions, with a cycle of destruction every 26 million years—which this hypothesis was intended to explain—was extracted by Raup and Sepkoski from Sepkoski's large paleontological database. But Nemesis has never been observed.

This Nemesis hypothesis attracted considerable attention at the time, but later it fell out of favor. It reappeared, however, in *Nature* on March 10, 2005, in a paper by Muller and his postdoctoral student Robert Rhode. They postulated a cycle of 62 million years (not 26 million!) for fossil extinctions based on a reevaluation of the now computerized and updated Chicago database.

Muller still seemed to favor an extraterrestrial explanation of the extinctions, whereas (according to media reports) Rhode thought there might be periodic activity within the Earth and a cyclic emission of flood basalts that could lead to extinctions. Be this as it may, the mathematical (Fourier) analysis of the numbers of known genera over time succeeded in extracting a strong periodic 62-million-year "signal" from the genera/time data. The curve also showed several sharp extinctions that might be associated with bolide impacts. The averaged plot of numbers of marine genera from the Cambrian to the end of the Tertiary showed an upward sweep that peaked in the Devonian and then declined to a minimum in the Triassic before climbing steadily toward the present. There appear, then, to have been two large periods of evolutionary expansion that seemingly correlated with the geochemical data discussed in Chapter 13 (Figure 13.5).

In discussing the paper in the same issue of *Nature,* James Kirchner and Anne Weil (2005) seemed disinclined to accept the astronomical explanation of the data but wondered whether the cause might have to do with the available size of the continental shelves (where most of the organisms would have lived and died). Or biodiversity itself might have its own natural swing—like a giant pendulum. "Excessive" diversity on a finite planet might, in itself, lead to a subsequent decline the numbers of species (genera?). Kirchner and Weil looked forward to "[c]lever modellers" getting to work on the problem and offering an essentially biological/ecological explanation. Or is it all driven by plate tectonics, eustasy, and geochemistry?

Where the debate will go next and whether such ideas will catch on or be forgotten I do not here speculate. An excellent survey of extinctions by Antony Hallam (2004) canvasses the various explanations that have been proposed to account for them (bolides, climate change, sea-level changes, marine "anoxia," volcanism, and human agencies) and concludes that only the end-Cretaceous extinctions seem to be directly related to a bolide, and even that case remains controversial.

Brief mention should also be made here of the *unique* appearance of a multitude of shelly life forms in the Cambrian—the so-called Cambrian explosion—following the billions of years of living organisms being little more than bacteria, algal mats, or soft-bodied creatures resembling jelly-fish in the late Precambrian. Various explanations have been proposed for the "explosion," and one, due to Paul Hoffman, that is attracting current attention is the occurrence of almost worldwide glaciation, followed by a "sudden" warming produced by the sudden escape of radiogenic heat that had been trapped under the ice ("Snowball Earth" hypothesis). This great disruption may have stimulated a huge evolutionary spurt (Hoffman et al. 1998; Walker 2003). If this suggestion achieves general acceptance, it will be anything but a cyclical theory. What next?

TIMELINE

5th c. B.C.	Xanthus reported marine shells found inland.
5th c. B.C.	Xenophenes: idea of hydrological cycle.
4th c. B.C.	Empedocles: four elements; studies of Etna; fires within a porous Earth.
4th c. B.C.	Herodotus: slowness of sedimentation; land and sea boundaries not fixed.
4th–3rd c. B.C.	Plato and Aristotle. Passages and cavities within Earth, with fires, winds, and waters. Aristotle's geocentric cosmos of infinite (?) age.
1st c. A.D.	Stoics (Seneca): cyclic interconversions of elements.
10th c.	"Brethren of Purity" encyclopedists in Basra: cyclic interchanges of land and sea.
14th c.	Buridan: cyclic interchanges of land and sea due to shifts of Earth's center of gravity. "Balance theory."
1644	Descartes: vortices; "mechanical" theory of the Earth.
1658	Ussher's estimate of 4004 B.C. for Creation date.
1667/1668, 1686/1687	Hooke on fossils, earthquakes, polar wandering, species changes, cycles of erosion/deposition, astronomical tests for pole wandering.
1669	Steno's *Prodromus*. Superposition and order of formation principles. Two cycles of Earth history. Fossils' origin understood.
1684	Burnet's *Sacred Theory of the Earth*.
1785, 1788	Hutton's cyclic theory of the Earth, unconformities, origin of granite.
1788	Siccar Point unconformity visited.
1797	Death of Hutton; birth of Lyell.

c. 1790–c. 1810	Emergence of geology as a recognizable independent science concerned with the Earth's history.
Early 1800s	Cuvier: methods for comparative anatomy enunciated and deployed.
1802	Lamarck's *Hydrogéologie*.
1812	Cuvier: *Recherches sur les ossemens fossiles*; geological "revolutions."
1815	Smith's geological map (*Delineation of the Strata of England and Wales*), with accompanying *Memoir*.
1816	Smith: *Strata Identified by Organized Fossils*.
1820s	Zenith of diluvial theory.
1828	Lyell's Etna visit. Compelling evidence for Earth's antiquity.
1829–1830	Élie de Beaumont's geotectonic theory adumbrated. Cooling Earth.
1830–1833	Lyell's *Principles of Geology*. Gradualism, actualism, non-directionalism; environment determines species' appearance or extinction. "Great Year."
1832	Whewell: "uniformitarianism" and "catastrophism." Terms coined.
1837–1840	Agassiz's glacial theory. *Eiszeit*.
1838	Schönbein: geochemistry proposed as a science.
1840s	Dana's ideas on cooling Earth. Permanence of continents and oceans.
1850s	Bischoff: chemical cycling studies.
1850s–1880s	Reconnaissance surveys of American West.
1855	Morlot: multiple glaciations.
1857	Hall: sediments deposited in sinking basins.
1870s	Dana: further ideas on mountain building.
1871	Dutton: isostasy.
1872–1876	*Challenger* expedition.
1875	Croll: *Climate and Time*. Astronomical theory of ice ages. Powell: base-level concept. Suess: *Entstehung der Alpen*. Lateral forces. *Vorland* and *Hinterland*.
1876	Gilbert: "grade" concept.
1878	De Geer: varves deposited annually, analogous to tree rings. First International Geological Congress.
1882	Penck: four glaciations recognized from river terraces to north of the Alps.
1883–1889	Suess: *Antlitz der Erde:* theory of cooling Earth. Eustasy (1888).
1889	Davis: landform cycles (youth, maturity, old age).
1894	Walther's facies law.
1897	Wiechert: *Mantel* proposed.
1898, 1909	Chamberlin: worldwide correlations (diastrophism causing eustatic changes).

1906	Ampferer: convection currents within Earth. Oldham: Earth's fluid core.
1911/1912	Wegener adumbrates continental drift theory.
1912	Udden: sedimentation cycles.
1913	Holmes: *Age of the Earth* (radiometric determinations). Expanded time scale.
1915	Wegener: *Entstehung der Kontinente und Ozeane*.
1917	Barrell: analysis of rhythmic sedimentation.
1920s	Vernadsky's biogeochemical studies.
1921	Krober: orogens and kratogens.
1924	Milanković: insolation curves, compatible with Penck's four glacial epochs.
1929	Holmes: convection currents to explain "drift" and geosyncline formation.
1929 (still ongoing)	Cyclothem studies.
1929	Matuyama: Earth's field reversals.
1931	Weller: diastrophic cycle.
1932	Wanless and Weller: the term "cyclothem" introduced.
1935	Wadati: inclined "earthquake surfaces."
1936	Wanless and Shepard: glacial-eustasy and cyclic sedimentation.
1939	Johnson: turbidites.
1941	Summation of Milanković's theory: *Kanon der Erdestrahlung*.
1947	Libby: carbon dating. Urey: oxygen isotope ratios as possible proxy for water temperatures.
1948	Kuenen: experimental sedimentology. Wheeler: time stratigraphy; Wheeler diagrams (later so called).
1948–1968	Nieuwenkamp: development of persedimentary model and sediment cycling.
1949	Sloss et al.: integrated facies analysis (forerunner of sequence stratigraphy).
1954	Benioff: dipping "earthquake surfaces."
1955	Emiliani: correlation of Milanković curves with oxygen isotope ratios for deep-sea cores.
1959	Vail's first sea-level curve.
1960s	Exxon's sequence stratigraphy program initiated. Geomagnetic polarity timescale developed.
1961	Dietz: seafloor spreading.
1962	Bouma sequences.
1963	Vine and Matthews: seafloor spreading and geomagnetic reversals.
1965	Wilson: "hot spots" and transform faults.
1966	Wilson: closing and reopening of Atlantic Ocean.

1971	Garrels and Mackenzie: *Evolution of Sedimentary Rocks.*
1976	Hays et al.: correlation of Milanković data, oxygen isotope data, and radiolarian types.
1977	Vail et al.: *Seismic Stratigraphy.* Vail curve published.
1979	Lovelock: *Gaia.*
1981	Mackenzie and Pigott: geochemical cycling and tectonics linked.
1981, 1984	Garrels and Lerman: coupling of sedimentary cycles for carbon and sulfur.
1984	Heckel: Milanković theory and cyclothems.
1984	"Nemesis" theory: cyclic catastrophes.
1992	"Snowball Earth" hypothesis.

INTERNATIONAL STRATIGRAPHIC CHART

Eon	Era	System/Period	Series/Epoch	Millions of years (Ma)	"Events"
Phanerozoic	*Cenozoic*	Neogene	Holocene/Recent	0–0.01	Modern humans
			Pleistocene	0.01–1.81	Ice ages; humans and Neanderthals
			Pliocene	1.81–5.33	*Homo erectus; Home habilis*
			Miocene	5.33–23.03	Australopithecenes; apes
		Paleogene	Oligocene	23.03–33.9	Global cooling; elephants; horses; grasslands
			Eocene	33.9–55.8	Modern mammals; whales; India/Asia collision
			Paleocene	55.8–66.5	Ended with high global temperatures and extinctions
	Mesozoic	Cretaceous	Upper	66.5–99.6	Chalk; dinosaur (etc.) extinctions at end-Cretaceous
			Lower	99.6–145.5	Deciduous trees; insect-pollinated flowering plants
		Jurassic	Upper	145.5–161.2	Flowering plants
			Middle	161.2–175.6	Feathered birds; marsupials
			Lower	175.6–199.6	Giant dinosaurs
		Triassic	Upper	199.6–228.0	Pangea break-up; Ginkgo; *Archeopteryx;* crocodiles
			Middle	228.0–245.0	Dinosaurs; mammals; cycads
			Lower	245.0–251.0	Thecodonts

(Continued)

Eon	Era	System/Period	Series/Epoch	Millions of years (Ma)	"Events"
Phanerozoic	*Paslaezoic*	Permian	Lopingian	251.0–260.4	Archosaurians; terminated by mass-extinctions
			Guadalupian	260.4–270.6	Glaciation; ichthyosaurs; plesiosaurs; snakes
			Cisuralian	270.6–299.0	Mammal-like reptiles
		Carboniferous	Pennsylvanian	299.0–318.1	Reptiles
			Mississipian	318.1–359.2	Conifers; amniotes
		Devonian	Upper	359.2–385.3	Amphibia; ferns
			Middle	385.3–397.5	Bony fish
			Lower	397.5–416.0	Mosses; liverworts
		Silurian	Pridoli	416.0–418.7	Ammonites
			Ludlow	418.7–422.9	Jawed fish with armor; sharks
			Wenlock	422.9–428.2	Insects
			Llandovery	428.2–443.7	Vascular plants; jawed fish
		Ordovician	Upper	443.7–460.9	Black graptolite shales common in Ordovician
			Middle	460.9–471.8	Bryozoans
			Lower	471.8–488.3	
		Cambrian	Upper	488.3–501.0	
			Middle	501.0–513.0	
			Lower	513.0–542.0	"Cambrian explosion:" hard-bodied animals, including chordates
Precambrian	*Proterozoic*	Neoproterozoic	Ediacaran	542–635	Macro soft-bodied organisms (Ediacara: Austrailian locality)
			Cryogenian	635–850	Global glaciation "Snowball Earth" (*Cryos* = ice)
			Tonian	850–1000	Further platform extension (*Tonas* = stretch)
		Mesoproterozoic	Stenian	1000–1200	Rodinia assembled; narrow metamorphic belts formed; first mantle convection? (*Stenos* = narrow
			Ectasian	1200–1400	Extension of platform covers (*Ectsis* = extension)
			Calymmian	1400–1600	Formation and expansion of platform covers on cratonized basements (*Calymma* = cover)
		Paleoproterozoic	Statherian	1600–1800	Nena; complex single-celled life (*Statheros* = stable, firm)
			Orosirian	1800–2050	Global orogeny; impacts; oxygen in atmosphere (*Orosira* = mountain range)
			Rhyacian	2050–2300	Photosynthesis; green algae (*Rhyax* = lava stream)
			Siderian	2300–2500	Oxygen-liberating algae; banded iron formations (*Sideros* = iron)

(Continued)

Eon	Era	System/Period	Series/Epoch	Millions of years (Ma)	"Events"
	Archean	Neoarchean	—	2500–2800	Slime moulds
		Mesoarchean	—	2800–3200	"Algal" mats; stromatolites
		Paleoarchean	—	3200–3600	Prokaryote bacteria; "extremophiles"
		Eoarchean (pre-3850 sometimes called Hadean)	—	>3600 Lower limit undefined	Oldest known rock (1999): 4031 Ma; oldest dated mineral (2001): 4408 Ma

Subdivisions and dates based on Gradstein, Ogg, and Smith (2004).

Origin of Earth by accretion of materials from solar nebula 4567 Ma.

N.B. All ages for the Precambrian are allocated according to radiometric determinations. Organisms mentioned in the "Events" column refer to their first appearance on Earth; most disappearances are not indicated.

GLOSSARY

Acidic rocks: Igneous rocks with 60 percent more silica and with 10 percent more free quartz.

Albite: A feldspar mineral.

Andesite: Volcanic rock containing such minerals as andesine, pyroxenes, hornblende, or brown mica. Found in the Andes.

Antecedent drainage: A drainage pattern that is independent of underlying geological structure; established prior to faulting, folding, or uplift, with rivers cutting down as land surface is elevated.

Aphelion: Point in Earth's orbit furthest from the Sun.

Asthenosphere: Plastic layer of the mantle, near its upper surface, immediately below the lithosphere.

Back-arc: A depression landward of a volcanic arc in a subduction zone, lined with sediment from the volcanic arc and the plate interior.

"Balance hypothesis": Supposed adjustments to Earth's shape following erosion and deposition that maintain its center of gravity at center of cosmos (Buridan and others).

Basalt: Dark, fine-grained, igneous rock (<53 percent silica).

Base level: Imagined plane surface underlying a landmass below which erosion cannot occur.

Base leveling: Erosion toward or down to a base level.

Basic rocks: Igneous rocks of relatively high iron, calcium, and magnesium content, with 45 to 53 percent silica.

Basin and range: Topography or landscape produced by block faulting, which has produced asymmetric mountain ranges and broad intervening basins.

Batholith: Large mass of igneous rock, usually granitic and emplaced at depth, which may be exposed by erosion.

Bedding plane: Planar surface separating one sediment layer from another, marking a minor depositional break.

Belemnite: Fossil cephalopod with a bullet-shaped hard part.

Biogeochemistry: Branch of geochemistry dealing with the effects of life on the distribution and fixation of elements in the biosphere.

Biostratigraphy: Stratigraphy based on the specification of rock units on the basis of fossils.

Bouma sequence: Succession of five different sediment types/structures found in turbidites.

Calcite/calcareous spar: Crystalline calcium carbonate.

Cambrian explosion: The sudden appearance of a variety of shelly organisms at the beginning of the Cambrian.

Carbon-14 dating: Method for dating carbon-containing materials according to the $^{14}C/^{12}C$ ratio. This changes over time because of radioactive decay of ^{14}C to ^{12}C at a known rate.

Catastrophism: The doctrine that Earth history has been interrupted from time to time by "catastrophes" (e.g., "mega-tsunamis" or impacts of objects from space).

Chlorite: A greenish, platy, low-grade metamorphic mineral.

Chronostratigraphy: Branch of stratigraphy concerned with determining the ages of strata and their time relations.

Comparative anatomy: Use of analogies from modern organisms to guide the reconstruction of extinct organisms.

Consequent drainage: Drainage pattern established on a land surface following earth movements or establishment of a volcano.

Continental drift: Supposed relative movements of continents, as envisaged by Wegener. See also "Mobilism" and "Plate tectonics."

Contraction theory: Ascribes major deformations of Earth's crust to cooling and contraction since its formation.

Country rock: Rock into which some igneous material is intruded.

Craton: Stable continental part of Earth's crust, no longer affected by mountain building.

Crust: Thin outer layer of the Earth.

Cyanobacteria: Bacteria with chlorophyll and capacity to photosynthesize (formerly called blue-green algae).

Cyclothem: Series of varied beds deposited during a cycle of sedimentation.

Davisian cycle: Landscape stages: "youth," "maturity," and "old age." Uplift produces "rejuvenation" and a renewed erosion cycle.

Diapirism: Piercing of rocks by upward movement of buoyant low-density materials.

Diastem: Small depositional break.

Diastrophism: Crustal movements caused by tectonic processes.

Diluvium: Assorted superficial deposits, formerly associated with Noachian Flood.

Directionalism: Assumption that the Earth's history has a direction.

Dollo's law: Evolution never reproduces new species exactly; evolution is irreversible.

Dolomite: Sedimentary rock: calcium and magnesium carbonates.

Downlap: Younger inclined beds thinning out and terminating down-dip at an underlying surface.

Drift: Superficial material, deposited by glaciers, supposedly carried by floating icebergs. See also "Diluvium."

Dyke: Tabular igneous intrusion that cuts across strata.

Dynamo hypothesis: Earth's magnetic field explained in terms of interacting electric currents and magnetic lines of force within the Earth.

Ecliptic: The plane defined by the Earth's orbit around the Sun.

Eclogite: Coarse-grained igneous rock of bulk chemical composition similar to basalt, composed chiefly of garnets and pyroxenes.

Enstatite: A pyroxene mineral found in basic and intermediate igneous rocks.

Epicontinental: Situated on the continental shelf.

Equinox: One of the two days each year when day and night are equal in length all over the Earth.

Erratics: Boulders made from rock displaced by agency of glaciers.

Eustasy: Worldwide changes in sea level due to tectonic movements or growth/decay of ice sheets.

Evaporites: Sedimentary rocks formed by the precipitation of salts from natural brines.

Facies: General appearance and characteristics of a rock, indicating its mode of origin.

Feldspar: Group of white or flesh-pink minerals: alumino-silicates of sodium, potassium, calcium, and so on.

"Fixism": Doctrine that continents have maintained their positions since their first formation.

Foliation: Laminated or fissile rock structure due to platy or tabular minerals, particularly micas, typically found in metamorphic rocks.

Foraminifera: Small single-celled marine organisms, with one- or many-chambered shells.

Fore-arc: Area on the subduction trench side of a line of volcanoes.

Gaia: Greek Earth goddess. Used to refer to the geocosm when regarded as a self-sustaining system, analogous to a living organism.

Geochemical cycling: Exchanges of elements between different geochemical reservoirs.

Geochemistry: Study of the chemistry of the Earth.

Geomagnetic reversals: Reversals in the direction of the Earth's magnetic field.

Geomagnetism: Magnetic phenomena of the Earth and its atmosphere; the study of such phenomena.

Geostrophic cycle: Term used to refer to the rock cycle. (See Figure 5.1)

Geosynclinal/geosyncline: Trough-like area of crustal down-warp in which great sediment thicknesses may accumulate or have accumulated.

Glacio-eustasy: Global sea-level changes caused by formation or melting of ice with the onset or termination of ice ages.

Gneiss: Coarse-grained metamorphic rock of granitic composition with alternating layers of granular or laminar minerals.

Gradualism: See "Uniformitarianism."

Granite: Coarse-grained igneous rock with quartz, feldspar(s), and mica as principal constituents.

Great Year: Period for stars to return to the same apparent positions in the sky. Movement due to precession. For Lyell: period of return of similar climatic conditions.

Greenalite: Earthy or greenish mineral, found in clays and serpentine.

Greywacke: Gray, gritty sandstone.

Guyot: Flat-topped submarine mountain.

Gypsum: Mineral: calcium sulfate

Igneous rock: Rock "formed by fire," that is, by solidification from a molten or partly molten state.

Inner core: The solid innermost part (iron) of the Earth.

Insolation: Exposure to Sun's radiation.

Island arc: Chain-like volcanic archipelago, lying on the continental side of an oceanic trench in a lithospheric "plate."

Isostasy: "Equal standing." Condition of equilibrium, as for a floating object, of portions of the lithosphere on underlying plastic asthenosphere.

Isotopes: Different kinds of the *same* chemical element, with different numbers of neutrons per atom but the same number of protons. Have different mass numbers.

Isotopic deviation (δ): Deviation of the ratios of two isotopes present in a sample from some standard ratio. (See p. 166)

Isotopic ratio: The ratio of two isotopes of the same element present in a sample.

Lithosphere: Rocks of the Earth's crust; also the upper brittle part of the mantle.

Lithostratigraphy: Stratigraphy based on the study of lithological characters of strata.

Magma: Molten material from within the Earth that forms igneous rocks on cooling and solidification.

Mantle: Large, rocky, solid (but semiplastic) part of the Earth's interior, in the region between its core and outer crust.

Mass spectrometer: Instrument for measuring the masses and relative concentrations of atoms and molecules in a sample, utilizing the action of a magnetic field on moving charged particles, which are separated according to their relative masses.

Mechanical philosophy: View that all physical changes could be explained in terms of motions and mechanical interactions of particles.

Meridian line: A great-circle line imagined on the Earth's surface, from pole to pole.

Metamorphic rock: Commonly foliated rock, formed from other rock by heat and/or pressure, which alter the texture and mineral composition.

Micas: Group of platy minerals: silicates of (chiefly) iron, sodium, potassium, magnesium.

Migma: Similar to magma but produced by penetration of hot fluids (magma or hydrothermal solutions) into country rock.

Milanković cycles: Cyclic climate changes arising from changes in insolation due to variations in the eccentricity of the Earth's orbit, inclination of its axis to the ecliptic, and precession.

Mobilism: Doctrine that continents' positions are not fixed relative to one another.

Normal magnetic polarity: Direction of Earth's magnetic field at present.

Obliquity of the ecliptic: Angle between Earth's axis of rotation and plane of the ecliptic.

Offlap: Withdrawal of the sea from land, indicated in the sedimentary record by the fact that each successive younger bed leaves a portion of the underlying older bed exposed.

Onlap: Submergence of a land surface by the sea such that the boundary of each sediment layer is transgressed by next overlying layer.

Oort cloud: Spherical cloud of comets at edge of the solar system.

Orbital forcing: Climatic changes caused by cyclic changes in the several components of the Earth's movements around the Sun.

Orogeny: Episode of crustal deformation/mountain building.

Outer core: Fluid part of Earth's interior, chiefly of iron.

Pangea: Original "supercontinent" envisaged by Wegener, which subsequently broke into separate continents.

Panthalassa: The "superocean" complementing Wegener's Pangea.

Paradigm: A mix of theory and practical techniques used in a successful piece of scientific research that serves as an exemplar for other related researches.

Pelite: Sediment made of finest-grained clay/mud particles.

Peneplain: Extensive plain: end product of erosion.

Peridotite: Ultrabasic rock, formed under high-pressure conditions, forming much of the mantle.

Perihelion: Point in the Earth's orbit nearest the Sun.

Persedimentary model: A model of geocycling according to which all rock materials exist in the form of sediments at some time.

Phanerozoic: Stratigraphic column from Cambrian to present.

Plate tectonics: Theory that there are large crustal "plates" on the Earth's surface, able to move laterally because of the asthenosphere's plasticity and the mantle's convection currents; also associated tectonic processes.

Platform: A part of a continent covered by gently tilted, most sedementary strata, underlain by basement rocks consolidated in earlier deformations and forming part of a craton.

Plume: Upward-rising current of buoyant material, rising through the Earth's mantle into its crust.

Pole wandering: Movements of polar positions produced by changes in distribution of material on or within the Earth.

Potassium/argon dating: Technique for determining ages of rocks/minerals, dependent on radioactive decay, at a known rate, of ^{40}K into ^{40}Ar.

Precession: Rotation of the axis of a spinning body acted on by some torque, such as frictional drag on a spinning top.

Precession of equinoxes: Slow change of direction in space of the Earth's axis such that the pole of the equator describes a circle around the pole of the ecliptic every 25,800 years.

Primitive rocks: Rocks formed when the Earth initially formed a solid crust, after its original formation (obs.).

Principle of superposition: In a sedimentary sequence, the lower layers were deposited first.

Pyrite: Brass-colored mineral (iron sulfide) with cubic crystals.

Quartz: Common mineral: silicon dioxide.

Quaternary: The fourth main subdivision of the Phanerozoic part of the Geologic Column (see p. 105), consisting of the Pleistocene (Glacial Epoch) and the Holocene (Recent), dating back to about 1.8 million years ago. (The term may eventually become obsolete.)

Radiolaria: Group of single-celled marine organisms with siliceous lattices and spicules.

Remanent magnetism: Component of a rock's magnetism, of fixed direction relative to the rock, which remains after applying a moderate external field. Indicative of the magnetism of the rock acquired at its formation.

Reversed magnetic polarity: Direction of Earth's magnetic field opposite to that at present.

Schist: A foliated, metamorphic rock, normally containing abundant mica.

Seafloor spreading: Process whereby, according to plate tectonics, basaltic matter rises from the Earth's interior along mid-oceanic ridges and spreads laterally, adding to the seafloor as continents move apart from one another.

Seismic stratigraphy: Study of stratigraphy by the interpretation of seismic data from explosions or vibrators at the surface.

Sequence: A packet of sediments (or a stratal package) produced in a sedimentary cycle, bounded above and below by unconformities or their correlative conformities, that can be regarded as a lithostratigraphic unit.

Sequence stratigraphy: A procedure for analyzing genetically related packages of sedimentary strata. It seeks to relate the deposition of strata across a developing basin to changes in sea level, elevation, subsidence, and concomitant sedimentation. It utilizes seismic stratigraphy as well as the usual stratigraphic tools. Also seeks to correlate sequences in different basins by their characteristic features and has produced graphs of sea-level changes over time.

Siderite: A brownish mineral and iron ore, found in clays and shales or as a precipitate.

Spectrum analysis: Mathematical technique for analyzing the amplitudes and wavelengths of the different components of a compound cyclic signal.

Stratigraphic column: Composite representation of the materials (strata) of all or part of geologic time. (See chart, p. 221–223)

Stratigraphy: The study of stratified rocks and their arrangement in space and time.

Strike: Direction of a line formed by the intersection of a rock surface with a horizontal plane. Perpendicular to the surface's dip.

Subduction: According to plate tectonic theory, the descent of material from the ocean floor below a continent into the Earth's interior near converging plate margins.

Suture zone: A belt of deformed rocks, marking the boundary between collided continents or island arcs.

Technoscience: Technology and science interpreted as mutually interacting components of a single discipline.

Tectonics: Branch of geology that deals with the broad structure of the upper crust of the Earth and its deformations and evolution.

Tethys: Former ocean, initially created by the rifting of Pangea, which occupied the area now marked by the Alpine–Himalayan orogenic belt; subsequently obliterated by the Alpine–Himalayan continental collision.

Thema (pl. themata): Concepts and preconceptions that guide scientists' imaginations and theorizing.

Till: Sediments deposited by glaciers (boulder clay).

Tillite: Lithified till.

Time stratigraphy: See "Chronostratigraphy."

Tourmaline: A group of variously colored minerals, often needle shaped, containing boron. Acquire electric charges on heating.

Transform faults: Strike-slip faults under an ocean, developed in association with the extrusion of magma from an oceanic ridge as the lavas adjust to the Earth's spherical surface.

Transformism: Belief that one species may change over time into another species.

Troilite: Mineral: iron sulfide.

Turbidite: Sediment or rock deposited by a turbidity current.

Turbidity current: Underwater current flowing swiftly downslope because of the weight of the sediment it carries.

Ultrabasic rocks: Igneous rocks with high proportion of iron and magnesium, no free quartz, and <45 percent silica.

Unconformity: Substantial break in the geologic record where a unit is overlain by another not next to it in the stratigraphic succession, usually marked by structural discordance (e.g., horizontal strata lying over inclined strata).

Uniformitarianism: The several doctrines that conditions on Earth remain approximately constant over time, that the laws of nature are constant, that past geological phenomena should be interpreted in terms of what can presently be observed, that geological history has no overall direction, and that geological processes and changes proceed gradually.

Vail curve: A curve showing changes in sea levels over time.

Varves: Thin layers of clay and silt representing annual deposits, usually in a lake formed by a retreating ice sheet.

Wadati–Benioff zones: Large fault zones where subduction is occurring, detectable by earthquakes generated at the fault planes.

Walther's law: A vertical facies sequence is the product of correlated, spatially adjacent, depositional environments.

Wilson cycle: Staged evolution of the opening and closing of an ocean basin, leading to the eventual "suturing" of two continental masses. (See p. 103)

SUGGESTIONS FOR FURTHER READING

Baxter, Stephen. 2003. *Revolutions in the Earth: James Hutton and the Age of the World.* London: Weidenfeld & Nicolson.
  A good semipopular account of Hutton's life and work.

Coe, Angela, ed. 2003. *The Sedimentary Record of Sea-Level Change.* Cambridge: Cambridge UP and Open UP.
  A beautifully illustrated exposition of sequence stratigraphy, primarily for geology students.

Chorley, Richard, Antony Dunn, and Robert Beckinsale. 1964. *The History of the Study of Landforms . . . Geomorphology before Davis.* London: Methuen and Wiley.
  The standard book on the early history of geomorphology, with emphasis on the United States.

Chorley, Richard, Robert Beckinsale, and Antony Dunn. 1973. *The History of the Study of Landforms . . . The Life and Work of William Morris Davis.* London: Methuen.
  Biography of Davis, with extensive discussion of his relations with the Pencks, father and son.

Dott, Robert, ed. 1992. *Eustasy . . .* Boulder, CO: Geological Society of America.
  A series of valuable essays on the history of eustasy studies, especially ideas about cyclothems.

Drake, Ellen 1996. *Restless Genius: Robert Hooke and his Earthly Thoughts.* New York: Oxford UP.
  A thorough examination of Hooke's geology, with an edition of his geological writings.

Glen, William. 1982. *The Road to Jaramillo: Critical Years of the Revolution in Earth Science.* Stanford, CA: Stanford UP.

A detailed but accessible account of the establishment of the geomagnetic time scale.

Gohau, Gabriel. 1991. *A History of Geology* . . . New Brunswick, NJ: Rutgers UP.

A valuable introductory account of the history of geology.

Gould, Stephen. 1987. *Time's Arrow Time's Cycle: Myth and Metaphor in the Discovery of Geological Time.* Cambridge, MA: Harvard UP.

Examination of the works of Burnet, Hutton, and Lyell, differentiating cyclic and linear views of Earth history.

Hallam, Tony. 2004. *Catastrophes and Lesser Calamities: The Causes of Mass Extinctions.* Oxford: Oxford UP.

A lucid discussion of the possible causes of mass extinctions: climate changes, volcanism, oxygen content of oceans, sea-level changes, and bolides.

Imbrie, John, and Katherine Imbrie. 1979. *Ice Ages: Solving the Mystery.* London: Macmillan.

An excellent account of the work leading to Milanković's rehabilitation and establishment of the astronomical theory of the Ice Ages.

Le Grand, Homer 1988. *Drifting Continents and Shifting Theories* . . . Cambridge: Cambridge UP.

A reliable account of the plate tectonics revolution from a philosophy of science perspective.

Oldroyd, David R. 1996. *Thinking about the Earth: A History of Ideas in Geology.* London: Athlone Press and Harvard UP.

A general history of geology, with coverage of some of the issues, such as geochemical cycling, treated in the present book, but in less detail.

Rudwick, Martin. 2005. *Bursting the Limits of Time: The Reconstruction of Geohistory in the Age of Revolution.* Chicago and London: Chicago UP.

A detailed study of the emergence of geology as a distinctive historical science around 1800. Emphasis on French contributions and the work of Cuvier.

Şengör, Celâl. 2003. *The Large-Wavelength Deformations of the Lithosphere: Materials for a History of the Evolution of Thought from the Earliest Times to Plate Tectonics.* Boulder, CO: Geological Society of America.

A work of massive but accessible scholarship, treating ideas about land elevation from antiquity to the first half of the twentieth century. Also accounts of early explorations in the United States and American developments of ideas about geosyncline theory and mountain building. Helpful coverage of Élie de Beaumont and other European writers on tectonics.

Westbroek, Peter. 1992. *Life as a Geological Force . . .* New York: Norton.

    A delightful book, written from a "Gaia perspective," about ideas on geochemical cycling and the role of living organisms in geological processes. Special attention given to Robert Garrels.

Wilson, Leonard. 1972. *Charles Lyell: The Years to 1841: The Revolution in Geology.* New Haven, CT: Yale UP.

    Standard account of Lyell's early career, when he was developing his "uniformitarian" ideas.

Wood, Robert. 1985. *The Dark Side of the Earth.* London: Allen & Unwin.

    A readable, imaginative account of the plate tectonics revolution, emphasizing its social aspects.

BIBLIOGRAPHY

ABBREVIATIONS

AAPG	American Association of Petroleum Geologists
AJS	*American Journal of Science*
BAAPG	*Bulletin of the American Association of Petroleum Geologists*
BGSA	*Bulletin of the Geological Society of America*
BJHS	*British Journal for the History of Science*
CRASP	*Comptes rendus de l'Académie des Sciences, Paris*
E	*Episodes*
ENPJ	*Edinburgh New Philosophical Journal*
G	*Geology*
GSA	Geological Society of America
JG	*Journal of Geology*
N	*Nature*
PM	*Philosophical Magazine*
PTRSL	*Philosophical Transactions of the Royal Society of London*
(Q)JGS	*(Quarterly) Journal of the Geological Society of London*
S	*Science*
SA	*Scientific American*
SEPM	Society for Sedimentary Geology
UP	University Press
USGSB	*U.S. Geological Survey Bulletin*

Adhémar, Joseph. 1842–1844. *Révolutions de la mer.* Paris: Carilian-Goery and Dalmont.

Agassiz, Louis. 1837. "Des glaciers, des moraines, et les blocs erratiques." *Bibliothèque universelle des sciences, belles-lettres, et arts, . . . à Genève. Parties des sciences* 12: 369–394.

Agassiz, Louis. 1838. "Upon Glaciers, Moraines, and Erratic Blocks." *ENPJ* 24: 364–383.

Agassiz, Louis. 1840. *Études sur les glaciers.* Neuchâtel: Jent and Gassman.

Albert of Saxony. 1492. *Alberti de saxonia in Aristotelis libros de c[a]elo & mundo.* Venice: Bonetus Locatellus (Library of Congress, Incunabula 1492.A4, 2nd ed.).

Ampferer, Otto. 1906. "Über das Bewegungsbild, von Faltengebirgen." *Jahrbuch der kaiserlich-königlichen geologischen Reichanstalt* 56: 539–622.

Argand, Émile. 1924. "La tectonique de l'Asie." *Comptes rendus de la XIIIe Congrès International de Géologie* 1, Part 5: 171–372.

Argand, Émile. 1977. *Tectonics of Asia.* New York: Hafner.

Aristotle. 1962. *Meteorologica.* London: Heinemann; Cambridge, MA: Harvard UP.

Aristotle. 1965. *On Sophistical Refutations; On Coming-to-be and Passing Away; On the Cosmos.* London: Heinemann; Cambridge, MA: Harvard UP.

Arrhenius, Gustaf. 1952. "Sediment Cores from the East [Equatorial] Pacific." *Reports of the Swedish Deep-Sea Expedition 1947–1948*, ed. Hans Pettersson. 5, Fasc. 1. Götteborg: Elandes Boktrycker Aktiebolag.

Barrell, Joseph. 1917. "Rhythms and the Measurement of Geological Time." *BGSA* 28: 745–904 and plates.

Bartholomew, Michael. 1976. "The Non-Progression of Non-Progression: Two Responses to Lyell's Doctrine." *BJHS* 6: 261–303.

Benioff, Hugo. 1954. "Orogenesis and Deep Crustal Structure—Additional Evidence from Seismology." *BGSA* 65: 385–400.

Berger, André. 1977. "Support for the Astronomical Theory of Climatic Change." *N* 269: 44–45.

Berger, André. 1978. *Numerical Values of the Elements of the Earth's Orbit from 5000000 YBP to 1000000 YBP.* Louvain: Université Catholíque de Louvain Institut d'Astronomie et de Géophysique George Lemaitre, Contribution 35.

Berger, André. 1988. "Milankovitch Theory and Climate." *Reviews of Geophysics* 26: 624–657.

Birch, Thomas. 1756–1757. *The History of the Royal Society of London for Improving of Natural Knowledge from the First Rise.* London: A Millar.

Birkett, Kirsten, and David Oldroyd. 1991. "Robert Hooke, Physico-Mythology, Knowledge of the World of the Ancients and Knowledge of the Ancient World." In *The Uses of Antiquity,* ed. Stephen Gaukroger, 145–170. Dordrecht: Kluwer.

Blackett, Paul. 1961. "Comparison of Ancient Climates with the Ancient Latitudes Deduced from Rock Magnetic Measurements." *Proceedings of the Royal Society* 263 (Ser. A): 1–30.

Blackwelder, Eliott. 1909. "The Valuation of Unconformities." *JG* 17: 289–299.

Bouma, Arnold. 1962. *Sedimentology of Some Flysch Deposits . . .* Amsterdam: Elsevier.

Broecker, Wallace, David Thurber, John Goddard, The-Lung Ku, Robley Matthews, and Kenneth Mesolella. 1968. "Milankovitch Hypothesis Supported by Precise Dating of Coral Reefs and Deep-sea Sediments." *S* 159: 297–300.

Broecker, Wallace, and Van Donk, Jan. 1970. "Insolation Changes, Ice Volumes, and the O¹⁸ Record in Deep-Sea Cores." *Reviews of Geophysics and Space Physics* 8: 169–198.

Brunhes, Bernard. 1906. "Recherches sur la direction d'aimentation des roches volcaniques." *Journal de physique théorique et appliqué* 8 (Ser. 4): 705–724.

Brunhes, Bernard, and Pierre David. 1901. "Sur la direction d'aimantation dans des couches d'argille transformée en brique par des coulées de lave." *CRASP* 133: 155–157.

Brunhes, Bernard, and Pierre David. 1903. "Sur la direction de l'aimantation permanante dans diverses roches volcaniques." *CRASP* 137: 975–977.

Buridani, Iohannis (Jean Buridan). 1942. *Quaestiones super libris quatuor de caelo et mundo,* ed. Ernest Moody. Cambridge, MA: Mediaeval Academy of America.

Burnet, Thomas. 1684. *The Theory of the Earth* . . . London: R. Norton (1st Latin ed., 1681).

Chamberlin, Thomas. 1898. "The Ulterior Basis of Time Divisions and the Classification of Geological History." *JG* 6: 449–462.

Chamberlin, Thomas. 1909. "Diastrophism as the Ultimate Basis of Correlation." *JG* 17: 685–693.

Chambers, Robert. 1853. "On the Glacial Phenomena in Scotland and Some Parts of England." *ENPJ* 54: 229–282.

Chorley, Richard, Robert Beckinsale, and Antony Dunn. 1973. *The History of the Study of Landforms* . . . *Vol. 2: The Life and Work of William Morris Davis.* London: Methuen.

Clarke, Frank. 1908. *The Data of Geochemistry.* Washington: USGSB 330.

Clarke, Frank. 1924. *The Data of Geochemistry.* Washington: USGSB 770.

Cordier, Louis. 1827. "Essai sur la température de l'intérieur de la terre." *Annales des mines* 270: 53–138.

Cox, Allan, Richard Doell, and Brent Dalrymple. 1963. "Geomagnetic Polarity Epochs and Pleistocene Geochronometry." *N* 198: 1049–1051.

Craig, Gordon, Donald McIntyre, and Charles Waterston, eds. 1978. *James Hutton's Theory of the Earth: The Lost Drawings.* Edinburgh: Scottish Academic Press.

Croll, James. 1864. On the Physical Cause of the Change of Climate during Geological Epochs. *PM* 28 (Ser. 4): 121–137.

Croll, James. 1866. "On the Excentricity of the Earth's Orbit." *PM* 31 (Ser. 4): 26–28.

Croll, James. 1867. "On the Excentricity of the Earth's Orbit, and its Physical Relation to the Glacial Epoch." *PM* 33 (Ser. 4): 119–131.

Croll, James. 1875. *Climate and Time in their Geological Relations* . . . New York: Appleton.

Croll, James. 1884. "On the Causes of the Mild Polar Climates." *PM* 18 (Ser. 5): 268–288.

Dana, James. 1846a. "On the Early Conditions of the Earth's Surface." *American Journal of Science and Arts* 2 (Ser. 2): 347–348.

Dana, James. 1846b. "Volcanoes of the Moon." *AJS* 2 (Ser. 2): 335–355.

Dana, James. 1847a. "Geological Results of the Earth's Contraction in Consequence of Cooling" *AJS* 3 (Ser. 2): 176–188.

Dana, James. 1847b. "Origin of the Grand Outline Features of the Earth's Crust" *AJS* 4 (Series 2): 381–398.

Dana, James. 1849. *Geology* . . . Philadelphia: Sherman.

Dana, James. 1856. "On the Plan of Development in the Geological History of North America, with a Map." *AJS* 22 (Ser. 2): 335–349.

Dana, James. 1862. *Manual of Geology:* . . . . Philadelphia: Bliss; London: Trübner, 1863.

Dana, James. 1866. "Observations on the Origin of Some of the Earth's Features." *AJS* 42: 205–211 and 252–253.

Dana, James. 1873a. "On the Origin of Mountains." *AJS* 5 (Ser. 3): 347–350.

Dana, James. 1873b. "On Some Results of the Earth's Contraction from Cooling, Including a Discussion of the Origin of Mountains, and the Nature of the Earth's Interior; Part I." *AJS* 5 (Ser. 3): 423–443.

Dana, James. 1873c. "On Some Results of the Earth's Contraction from Cooling; Part II, The Condition of the Earth's Interior, and the Connection of the Facts with Mountain-Making; Part III, Metamorphism" *AJS* 6 (Ser. 3): 423–443; *AJS* 6 (Ser. 3): 6–14.

Dana, James. 1873d. "On Some Results of the Earth's Contraction from Cooling; Part IV, Igneous Ejections, Volcanoes." *AJS* 6 (Ser. 3): 104–115.

Dana, James. 1873e. "On Some Results of the Earth's Contraction from Cooling; Part V, Formation of Continental Plateaus and Oceanic Depressions." *AJS* 6 (Ser. 3): 161–172.

Darwin, Charles. 1859. *On the Origin of Species* . . . London: Murray.

Davis, Marc, Piet Hut, and Richard Muller. 1984. "Extinctions of Species by Periodic Comet Showers." *N* 308: 715–717.

Davis, William. 1883. "The Origin of Cross-Valleys." *S* 1; 325–327, 356–357.

Davis, William. 1889a. "A Pirate-River." *S* 13: 108–109.

Davis, William. 1889b. "The Rivers and Valleys of Pennsylvania." *National Geographic Magazine,* 1: 183–225.

Davis, William. 1897. "Current Notes on Physiography." *S* 6 (n.s.): 22–24.

Davis, William. 1932. "Piedmont Benchlands and the Primärrumpfe [initial peneplain]." *BGSA* 43: 399–440.

De Geer, Gerhard. 1912. "A Geochronology of the Last 12,000 Years." *Compte rendu de la XI:e Session du Congrès Géologique International (Stockholm 1910),* Fasc. 1: 241–253.

De Geer, Gerhard. 1927. "Geochronology Based on Solar Radiation." *S* 56: 000.

De Geer, Gerhard. 1940. "Geochronologica Suecica Principles." In *Kungliga Svenska Vetenskapsakademiens Handlingar.* Ser. 3, 18 (8): 1–367.

De Graciansky, Pierre-Charles, Jan Hardenbol, Theirry Jacquin, and Peter Vail, eds. 1998. *Mesozoic and Cenozoic Stratigraphy.* Tulsa, OK: SEPM.

Dean, Dennis. 1992. *James Hutton and the History of Geology.* Ithaca, NY: Cornell UP.

Delesse, Achille. 1849. "Sur le magnétisme polaire dans les minéraux et dans les roches." *Annales de chimique et de physiques* 25, 194–209.

Descartes, René. 1644/1656. *Renaté Des-Cartes Principia philosophiae.* Amsterdam: Ludovicum Elzevir.

Dewey, John, and Walter Pitman. 1998. "Sea-Level Changes: Mechanisms, Magnitudes and Rates." In *Paleographic Evolution and Non-Glacial Eustasy,* ed. J. Pindell and C. Drake, 1–16. Tulsa, OK: SEPM, Special Publication No. 58.

Dickinson, William. 2003. "The Place and Power of Myth in Geoscience . . ." *AJS* 303: 856–864.

Dietz, Robert. 1961. "Continental Ocean Basin Evolution by Spreading of the Sea Floor." *Nature* 190: 854–857.

Donovan, Arthur, and Joseph Prentiss. 1980. *James Hutton's Medical Dissertation. Transactions of the American Philosophical Society* 70: Part 6.

Dott, Robert, Jr. ed. 1992. *Eustasy: The Historical Ups and Downs of a Major Geological Concept.* Boulder, CO: GSA.

Dott, Robert, Jr., 1997. "James Dwight Dana's Old Tectonics—Global Contraction under Divine Direction." *AJS* 297: 283–311.

Drake, Ellen. 1996. *Restless Genius: Robert Hooke and His Earthly Thoughts.* New York: Oxford UP.

Du Toit, Alexander. 1937. *Our Wandering Continents . . .* Edinburgh: Oliver and Boyd.

Duhem, Pierre. 1958. *Le système du monde . . .* 9. Paris: Hermann.

Dunham Kingsley. 1950. "Lower Carboniferous Sedimentation in the Northern Pennines (England)." *Reports of the International Geological Congress for 1948* 4: 46–62.

Dutch, Steven. 1999. "History of Global Plate Motions." www.uwgb.edu/dutchs/platetec/plhist94.htm.

Dutton, Clarence. 1882. *Tertiary History of the Grand Cañon District.* Washington, DC: Government Printing Office.

Dutton, Clarence. 1892. "On Some of the Greater Problems of Physical Geology." *Bulletin of the Philosophical Society of Washington* 11: 51–64.

Duval, Bernard, Carlos Cramez, and Peter Vail. 1998. "Stratigraphic Cycles and Major Marine Source Rocks." In *Mesozoic and Cenozoic Sequence Stratigraphy of European Basins,* ed. De Craciansky, Pierre, Jan Hardenbol, Jaquin Thierry, and Peter Vail, 43–51. Tulsa, OK: SEPM.

Eighteenth International Geological Congress. 1950. "Recommendations of Commission Appointed to Advise on the Definition of the Pliocene–Pleistocene Boundary." *Report of the Eighteenth Session,* Part 9 (Section H): 6.

Einsele, Gerhard, and Werner Ricken. 1991. "Introductory Remarks." In *Cycles and Events in Stratigraphy,* ed. Gerhard Einsele, Werner Ricken, and Adolf Seilacher, 611–616. Berlin: Springer.

Eliade, Mircea. 1959. *The Myth of the Eternal Return.* New York: Harper.

Élie de Beaumont, Léonce. 1829–1830. "Recherches sur quelques unes des révolutions de la surface du globe." *Annales des sciences naturelles* 18: 5–25, 284–416; 19: 5–99, 177–240.

Élie de Beaumont, Léonce. 1831. "Researches on some of the Revolutions Which Have Taken Place on the Surface of the Globe . . . " *PM* 10 (n.s.): 241–264.

Élie de Beaumont, Léonce. 1850. "Extrait d'une lettre adressé à Prévost." *CRASP* 21: 501–505.

Élie de Beaumont, Léonce. 1852. *Notice sur les systèmes de montagnes.* 3 vols. Paris: Bertrand.

Ellenberger, François. 1996. *History of Geology,* 1. Rotterdam: Balkema.

Emiliani, Cesare. 1955. "Pleistocene Temperatures." *JG* 63: 538–578.

Emiliani, Cesare. 1966. "Paleotemperature Analysis of Caribbean Cores P6304–8 and P6304–9 and a Generalized Temperature Curve for the Past 425,000 Years." *JG* 74: 109–126.

Ferm, John C. 1975. "Pennsylvanian Cyclothems of the Appalachian Plateau, a Retrospective View." *USGS Professional Paper* 853: 57–64.

Forel, François A. 1885. "Les ravins sous-lacustrines des fleuves glaciares." *CRASP* 101: 725–728.

Francoeur, Louis-Benjamin. 1843. *Connaissance des temps ou des mouvements célestes, à l'usage des astronomes et des navigateurs.* Paris: Bachelier.

Frankel, Charles. 1999. *The End of the Dinosaurs: Chicxulub Crater and Mass Extinctions.* Cambridge: Cambridge UP.

Fritscher, Bernhard. 2002. "Alfred Wegener's 'The Origin of Continents,' 1912." *E* 25: 100–106.

Fuller, John. 2005. "A Date to Remember: 4004 BC." *Earth Sciences History* 24: 5–14.

Garrels, Robert, and Abraham Lerman. 1981. "Phanerozoic Cycles of Sedimentary Carbon and Sulfur." *Proceedings of the National Academy of Sciences: Physical Series* 78: 4652–4656.

Garrels, Robert, and Abraham Lerman. 1984. "Coupling of the Sedimentary Sulfur and Carbon Cycles An Improved Model." *AJS* 284: 989–1007.

Garrels, Robert, and Fred Mackenzie. 1971. *Evolution of Sedimentary Rocks.* New York: Norton.

Garrels, Robert, and Edward Perry. 1974. "Cycling of Carbon, Sulfur, and Oxygen through Geologic Time." In *The Sea: Volume 5, Marine Chemistry,* ed. Edward Goldberg, 303–336. New York: Wiley.

Geikie, Archibald. 1962. *The Founders of Geology.* New York: Dover (1st ed., 1897).

Geikie, James. 1874. *The Great Ice Age and Its Relation to the Antiquity of Man.* New York: Appleton.

Geikie, James. 1881. *Prehistoric Europe . . .* London: Stanford.

Gerstner, Patsy. 1968. "James Hutton's 'Theory of the Earth' and His Theory of Matter." *Isis* 59: 26–31.

Gerstner, Patsy. 1971. "The Reaction to James Hutton's Use of Heat as a Geological Agent." *BJHS* 5: 353–362.

Gignoux, Maurice. 1910. "Sur la classification du Pliocène et du Quaternaire de l'Italie du Sud." *CRASP* 150: 841–844.

Gilbert, Grove. 1876. "The Colorado Plateau Province as a Field for Geological Study." *AJS* 12 (Ser. 3): 6–24, 85–103.

Gilbert, Grove. 1877. *Report on the Geology of the Henry Mountains.* Washington, DC: Government Printing Office.

Gilbert, Grove. 1895. "Sedimentary Measurement of Cretaceous Time." *JG* 3: 121–127.

Glen, William. 1982. *The Road to Jaramillo . . .* Stanford, CA: Stanford UP.

Goldschmidt, Victor. 1958. *Geochemistry.* Oxford: Clarendon Press.

Gould, Stephen. 1987. *Time's Arrow Time's Cycle . . .* Cambridge, MA: Harvard UP.

Gould, Stephen. 1999. *Leonardo's Mountain of Clams and Diet of Worms.* London: Vintage.

Grabau, Amadeus. 1936. "Oscillation or Punctuation." *International Geological Congress: Report of the XVI Session. Washington 1933.* 1: 539–553.

Grabau, Amadeus. 1978. *The Rhythm of the Ages.* Huntington, NY: Krieger (1st ed., 1940).

Gradstein, Felix, James Ogg, and Alan Smith, eds. 2004. *A Geologic Time-Scale: 2004.* Cambridge: Cambridge UP.

Gradstein, Felix, James Ogg, Alan Smith, Wouter Bleeker, and Lucas Lourens. 2004. "A New Geological Time-Scale, with Special Reference to Precambrian and Neogene." *E* 27: 83–100.

Grant, Edward. 1974. *A Source Book in Medieval Science.* Cambridge, MA: Harvard UP.

Gregor, Bryan. 1966. *The Geochemical Behaviour of Sodium . . .* Amsterdam: Royal Netherlands Academy of Sciences.

Gregor, Bryan. 1988. "Prologue: Cyclic Processes in Geology, a Historical Sketch." In *Chemical Cycles in the Evolution of the Earth,* ed. Bryan Gregor, Robert Garrels, Fred Mackenzie, and Barry Maynard, 5–16. New York: Wiley.

Gregor, Bryan. 1992. "Some Ideas on the Rock Cycle: 1788–1988." *Geochimica et cosmochimica acta* 56: 2993–3000.

Guthrie, William. 1965. *A History of Greek Philosophy* 2. Cambridge: Cambridge UP.

Haarmann, Erich. 1930. *Die Oszillationstheorie; eine Erklärung der Krustenbewegungen von Erde und Mond.* Stuttgart: Enke.

Haber, Francis. 1959. *The Age of the World: Moses to Darwin.* Baltimore: Johns Hopkins UP.

Hall, James. 1859. "Introduction." In *Natural History of New York. Part 6. Palaeontology of New York, Vol. III, Part I.* Albany: New York State Geological Survey, van Benthuysen, 1–96.

Hall, James. 1860. "On the Formation of Mountain Ranges." *Canadian Journal of Industry, Science and Art* 5 (n.s.): 542–544.

Hall, James, and Josiah Whitney. 1858. "General Geology." In *Report on the Geological Survey of the State of Iowa, Vol. I, Part I.* Des Moines: State of Iowa, van Benthuysen, 35–44.

Hallam, Anthony. 1973. *A Revolution in the Earth Sciences* . . . Oxford: Clarendon Press.

Hallam, Anthony. 2004. *Catastrophes and Lesser Calamities* . . . Oxford: Oxford UP.

Halley, Edmund. 1715. "A Short Account of the Cause of Saltiness of the Oceans, . . . with a Proposal by Help Thereof, to Discover the Age of the World." *PTRSL* 29: 296–300.

Haq, Bilal, Jan Hardenbol, and Peter Vail. 1987. "Chronology of Fluctuating Sea Levels since the Triassic." *S* 235: 1156–1167.

Haq, Bilal, Jan Hardenbol, and Peter Vail. 1988. "Mesozoic and Cenozoic Chronostratigraphy and Cycles of Sea-Level Change." In *Sea-Level Changes* . . . , ed. Cheryl Wilgus, Henry Posamentier, Charles Ross, and Christopher Kendall, 71–108. Tulsa, OK: SEPM.

Harvey, William. 1628. *De motu cordis et sanguinis in animalibus.* Frankfurt: Sumptibus Guilielmi Fitzer.

Hays, James, and William Berggren. 1971. "Quaternary Boundaries and Correlations." In *The Micropaleontology of Oceans,* ed. Brian Funnell and William Riedel, 669–691. Cambridge: Cambridge UP.

Hays, James, John Imbrie, and Nicholas Shackleton. 1976. "Variations in the Earth's Orbit: Pacemaker of the Ice Ages." *S* 194: 1121–1133.

Hays, James, and Neil Opdyke. 1967. "Antarctic Radiolaria, Magnetic Reversals, and Climatic Change." *S* 158: 1001–1011.

Heckel, Philip. 1977. "Origin of Phosphatic Black Shale Facies in Pennsylvanian Cyclothems of Mid-Continent North America." *BAAPG* 61: 1045–1068.

Heckel, Philip. 1984. "Changing Concepts of Midcontinent Pennsylvanian Cyclothems, North America." In *Neuvième Congrès International de Stratigraphie et de Géologie du Carbonifère* . . . *1979: Compte rendu.* Carbondale: Southern Illinois UP. 3: 535–553.

Heckel, Philip. 1986. "Sea-Level Curve for Pennsylvanian Eustatic Marine Transgressive Depositional Cycles along Midcontinent Outcrop Belt, North America." *Geology* 14: 330–334.

Heckel, Philip. 1994. "Evaluation of Evidence Glacio-Eustatic Control over Marine Pennsylvanian Cyclothems in North America and Consideration of Possible Tectonic Effects." *Tectonic and Eustatic Controls on Sedimentary Cycles, SEPM Concepts in Sedimentology* 4: 65–87.

Heckel, Philip. 1995. "Glacial–Eustatic Base-Level—Climatic Model for Late Middle to Late Pennsylvanian Coal–Bed Formation in the Appalachian Basin." *Journal of Sedimentary Research* B65: 348–386.

Heidel, William. 1921. "Anaximander's Book, the Earliest Known Geographical Treatise." *Proceedings of the American Academy of Arts and Sciences* 56: 239–288.

Herschel, John. 1826–1833. "On the Astronomical Causes Which May Influence Geological Phenomena." *Proceedings of the Geological Society of London* 1: 244–245 (read 1830).

Herschel, John. 1837. "Extracts from Letters to Roderick Murchison and Charles Lyell." *The Ninth Bridgewater Treatise*, by Charles Babbage. London: Murray, 225–247.

Hess, Harry. 1962. "History of Ocean Basins." In *Petrologic Studies*, ed. Albert Engel, Harold James, and Benjamin Leonard, 599–620. New York: GSA.

Hinnov, Linda. 2000. "New Perspectives on Orbitally Forced Stratigraphy." *Annual Review of Earth and Planetary Sciences* 28: 419–475.

Hinnov, Linda. 2004. "Astronomical Signals from the Pre-Cambrian." In *Milutin Milankovitch Anniversary Symposium . . . 2004 Abstracts*, 51–73. Belgrade: Serbian Academy of Arts and Sciences.

Hoffman, Kenneth. 1988. "Ancient Magnetic Reversals: Clues to the Geodynamo." *SA* 258 (May): 50–59.

Hoffman, Paul, A. J. Kaufmann, Galen Halverson, and Daniel Schrag. 1998. "A Neoproterozoic Snowball Earth." *S* 281: 1342–1346.

Holmes, Arthur. 1929. "Radioactivity and Earth Movements." *Transactions of the Geological Society of Glasgow* 18: 559–606.

Holmes, Arthur. 1944. *Principles of Physical Geology*. London: Nelson (2nd ed., 1965).

Holser, William, Manfred Schidlowski, Fred Mackenzie, and Barry Maynard. 1988. "Geochemical Cycles of Carbon and Sulfur." In *Chemical Cycles in the Evolution of the Earth*, ed. Bryan Gregor, Robert Garrels, Fred Mackenzie, and Barry Maynard, 105–173. New York: Wiley.

Holton, Gerald. 1978. *The Scientific Imagination: Case Studies*. Cambridge: Cambridge UP.

Hooke, Robert. 1705. *Lectures and Discourses of Earthquakes . . . In Posthumous Works of Robert Hooke*, ed. Richard Waller, 279–450. London: Smith and Walford.

Hunt, T. Sterry. 1875. *Chemical and Geological Essays*. Boston: Osgood.

Hutton, James. 1788. "Theory of the Earth; Or an Investigation of the Laws Observable in the Composition, Dissolution, and Restoration of Land upon the Globe." *Transactions of the Royal Society of Edinburgh* 1: 209–304.

Hutton, James. 1792. *Dissertations on Different Subjects in Natural Philosophy*. Edinburgh: Cadell and Davies.

Hutton, James. 1794a. *A Dissertation upon the Philosophy of Light, Heat, and Fire*. Edinburgh: Strahan and Cadell.

Hutton, James. 1794b. "Observations on Granite." *Transactions of the Royal Society of Edinburgh* 3: 77–85.

Hutton, James. 1795. *Theory of the Earth . . .* 2 vols. Edinburgh: Creech; London: Cadell, Junior and Davies.

Hutton, James. 1899. *Theory of the Earth . . . Vol. III*, ed. Archibald Geikie. London: The Geological Society.

Imbrie, John, and Katherine Imbrie. 1979. *Ice Ages: Solving the Mystery*. London: Macmillan.

Jaki, Stanley. 1974. *Science and Creation . . .* Edinburgh: Scottish Academic Press.

Johnson, Douglas. 1939. *The Origin of Submarine Canyons* . . . New York: Columbia UP.

Joly, John, 1899. "An Estimate of the Geological Age of the Earth." *Transactions of the Royal Dublin Society* 7: 23–66.

Joly, John. 1909. *Radioactivity and Geology* . . . London: Constable.

Joly, John. 1930. *The Surface-History of the Earth.* Oxford: Clarendon Press.

Jukes, Joseph. 1862. "On the Mode of Formation of Some of the River-Valleys in the South of Ireland." *QJGS* 18: 378–403.

Kent, Dennis. 1999. "Orbital Tuning of Geomagnetic Polarity Time-Scales." *PTRSL,* Ser. A 357: 1995–2007.

King, William, and Kenneth Oakley. 1950. "Report of the Temporary Commission on the Pliocene–Pleistocene Boundary." In *18th International Geological Congress, Great Britain, 1948, Part 1,* 213–214. London.

Kirchner, James, and Anne Weil. 2005. "Fossils Make Waves." *N* 434: 147–148.

Klein, George. 1990. "Pennsylvanian Time-Scales and Cyclic Periods." *G* 18: 455–457.

Klein, George, and Debra Willard. 1989. "Origin of the Pennsylvanian Coal-Bearing Cyclothems of North America." *G* 17: 152–155.

Köppen, Wladimir, and Alfred Wegener. 1924. *Die Klimate der geologischen Vorzeit.* Berlin: Borntraeger.

Kuenen, Philip. 1948 (1950). "Turbidity Currents of High Density." *Proceedings of the 18th International Geological Congress, London, 1948* 8: 44–52.

Kuenen, Philip, and Carlo Migliorini. 1950. "Turbidity Currents as a Cause of Graded Bedding." *JG* 58: 91–127.

Kuhn, Thomas. 1962. *The Structure of Scientific Revolutions.* Chicago: University of Chicago Press.

Lamarck, Jean-Baptiste. 1802. *Recherches sur l'organisation des corps vivans* . . . Paris: Maillard.

Lamarck, Jean-Baptiste. Year 10 [1802]. *Hydrogéologie* . . . Paris: Agasse and Maillard.

Lamarck, Jean-Baptiste. 1809. *Philosophie zoologique* . . . 2 vols. Paris: Dentu.

Lamarck, Jean-Baptiste. 1815–1822. *Histoire naturelle des animaux sans vertèbres.* 7 vols. Paris: Verdière.

Lamarck, Jean-Baptiste. 1964. *Hydrogeology* . . . Urbana: University of Illinois Press.

Latour, Bruno. 1987. *Science in Action* . . . Milton Keynes: Open UP.

Le Grand, Homer. 1988. *Drifting Continents and Shifting Theories* . . . Cambridge: Cambridge UP.

Le Grand, Homer, and William Glen. 1993. "Choke-Holds, Radiolarian Cherts, and Davy Jones's Locker." *Perspectives on Science* 1: 24–67.

Leibniz, Gottfried. 1749. *De prima facie tellvris et antiqvissimae historiae vestigiis in ipsis natvrae monvmentis dissertation.* Göttingen: I. Schmidii.

Leibniz, Gottfried. 1993. *Protogaea* . . . Toulouse: Presses Universitaires du Mirail.

Leonardo da Vinci. 2000. *The Codex Leicester.* Sydney: Powerhouse Museum.

Lewis, Cherry. 2000. *The Dating Game . . .* Cambridge: Cambridge UP.

Libby, Willard. 1952. *Radiocarbon Dating.* Chicago: University of Chicago Press.

Locke, Jim. 2000. www.marin.cc.ca.us/~jim/photos/france/peiracava.html.

Lovelock, James. 1979. *Gaia . . .* Cambridge: Cambridge UP.

Lovelock, James. 1988. *The Ages of Gaia . . .* Oxford: Oxford UP.

Lyell, Charles. 1830. *Principles of Geology . . .* Vol. 1. London: John Murray.

Lyell, Charles. 1851. "Anniversary Address of the President." *QJGS* 7: xxv–lxxvi.

Lyell, C. 1863. *The Antiquity of Man . . .* London: Murray

Lyell, Charles. 1881. *Life Letters and Journals . . .* London: Murray.

Lyell, Charles. 1997. *Principles of Geology,* edited with an introduction by James Secord. London: Penguin Books (1st ed., 3 vols., 1830–1833).

Mackenzie, Fred, and John Pigott. 1981. "Tectonic Control of the Phanerozoic Sedimentary Rock Cycling." *JGS* 138: 183–196.

Macrobius, Ambrosius. 1952. *Commentary on the Dream of Scipio . . .* New York: Columbia UP.

Marvin, Ursula. 1973. *Continental Drift . . .* Washington, DC: Smithsonian Institution.

Mason, Ronald, and Arthur Raff. 1961. "Magnetic Survey off the Coast of North America, 32° N. Latitude to 52° N. Latitude." *BGSA* 72: 1267–1270.

Matuyama, Mononori. 1929. "On the Direction of Magnetisation of Basalt in Japan, Tyôsen and Manchuria." *Proceedings of the Imperial Academy of Japan* 5: 203–205.

McMenamin, Mark, and Dianna McMenamin. 1990. *The Emergence of Animals: The Cambrian Breakthrough.* New York: Columbia UP.

Meade, Mellard. 1877. "President's Address." *Proceedings of the Liverpool Geological Society* 3: 211–235.

Menard, William. 1986. *The Ocean of Truth . . .* Princeton, NJ: Princeton UP.

Miall, Andrew. 1990. *Principles of Sedimentary Basin Analysis.* 2nd ed. New York: Springer-Verlag.

Miall, Andrew. 1992. "Exxon Global Chart: An Event for Every Occasion?" *G* 20: 787–790.

Miall, Andrew. 1994. "Sequence Stratigraphy and Chronostratigraphy: Problems of Definition and Precision in Correlation, and their Implications for Global Eustasy." *Geoscience Canada* 21: 1–26.

Miall, Andrew. 1995. "Whither Stratigraphy?" *Sedimentary Geology* 100: 5–20.

Miall, Andrew, and Charlene Miall. 2001. "Sequence Stratigraphy as a Scientific Enterprise: The Evolution and Persistence of Conflicting Paradigms." *Earth-Science Reviews* 54: 321–348.

Miall, Andrew, and Charlene Miall. 2002. "The Exxon Factor: The Roles of Corporate and Academic Science in the Emergence and Legitimation of a New Global Model of Sequence Stratigraphy." *Sociological Quarterly* 43: 307–334.

Miall, Andrew, and Charlene Miall. 2004a. "Empiricism and Model-Building in Stratigraphy: Around the Hermeneutic Circle in the Pursuit of Stratigraphic Correlation." *Stratigraphy* 1: 27–46.

Miall, Andrew, and Charlene Miall. 2004b. "Empiricism and Model-Building in Stratigraphy: The Historical Roots of Present-Day Practices." *Stratigraphy* 1: 1–25.

Milanković, Milutin. 1920. *Théorie mathématique des phénomènes thermiques produits par la radiation solaire.* Paris: Gauthier-Villars.

Milanković, Milutin. 1930. *Mathematische Klimatlehre und Astronomische Theorie der Klimaschwankungen: Köppen Geigersches Handbuch der Klimatologie* 1 (Part A), Berlin.

Milanković, Milutin. 1938. "Neue Ergebnisse der astronomische Theorie der Klimaschwankungen." *Bulletin de l'Académie Royale Serbe des Sciences Mathématiques et Naturelles: A. Sciences mathématiques et physiques* 4.

Milanković, Milutin. 1998. *Canon of Insolation and the Ice-Age Problem.* Belgrade: Zavod Udžbenike i Nastavna Sredstva (first published in German by Serbian Royal Academy, 1941, as *Kanon der Erdbestrahlung und sein Anwendung für das Eiszeitproblem*).

Mitchum, Robert. 2003. Penrose Medal Citation. http://www.geosociety.org/aboutus/awards/03speeches/penrose.htm.

Moore, Derek. 1958. "The Yoredale Series of Upper Wensleydale and Adjacent Parts of North-East Yorkshire." *Proceedings of the Yorkshire Geological Society* 31: 91–146.

Moore, Derek. 1959. "Role of Deltas in the Formation of Some British Lower Carboniferous Cyclothems." *JG* 67: 522–539.

Moore, Raymond. 1929. "Environment of Pennsylvanian Cyclothems, North America." *BAAPG* 13: 159–487.

Moore, Raymond. 1931. "Pennsylvanian Cycles in the Northern Mid-Continent Region." *Bulletin of the Geological Survey of Illinois* 60: 247–257.

Moore, Raymond. 1936. "Stratigraphic Classification of the Pennsylvanian Rocks of Kansas." *Bulletin of the Geological Survey of Kansas* 22.

Moore, Raymond, et al. 1944. "Correlation of Pennsylvanian Formations of North America." *BGSA* 55: 657–706.

Morley, Lawrence, and André Larochelle. 1964. "Palaeomagnetism as a Means of Dating Geological Events." In *Geochronology in Canada . . .* ed. Freleigh Fitz Osborne, 39–51. Toronto: Royal Society of Canada and Toronto UP.

Morlot, Adolph. 1855. "On the Post-Tertiary and Quaternary Formations of Switzerland." *ENPJ* 2 (n.s.): 14–29.

Murphy, Brendan, and Damian Nance. 1991. "Supercontinental Model for the Contrasting Character of the Late Proterozoic Orogenic Belts." *G* 19: 469–472.

Murphy, Brendan, and Damian Nance. 1992. "Mountain Belts and the Supercontinental Cycle." *SA* 266 (4): 34–41.

Murphy, Joseph. 1869. "On the Nature and Cause of the Glacial Climate." *QJGS* 32: 350–356.

Murphy, Joseph. 1876. "The Glacial Climate and the Polar Icecap." *QJGS* 32: 400–406.

Neal, Jack, David Risch, and Peter Vail. 1993. "Sequence Stratigraphy—A Global Theory for Global Success." *Oilfield Review*, January, 51–62. http://www.oilfield.slb-com/media/services/oilfieldreview/ors93/0193/p51_62.pdf.

Nieuwenkamp, Willem. 1948. "Geochemistry of Sodium." *Report of the International Geological Congress, 18th Session, London, 1948* 2: 96–100.

Nieuwenkamp, Willem. 1956. "Géochimie classique et transformiste." *Bulletin de la Société Géologique de France*, Ser. 6, 6: 407–429.

Nieuwenkamp, Willem. 1968. "Oceanic and Continental Basalts in the Geochemical Cycle." *Geologische Rundschau* 57: 362–372.

Oldham, Richard. 1906. "The Constitution of the Earth as Revealed by Earthquakes." *QJGS* 62: 456–475.

Oldroyd, David. 1972. "Robert Hooke's Methodology of Science as Exemplified in His 'Discourse of Earthquakes.'" *BJHS* 6: 109–130.

Oldroyd, David. 1989. "Geological Controversy in the Seventeenth Century: 'Hooke *vs.* Wallis' and its Aftermath." In *Robert Hooke: New Studies*, ed. Michael Hunter and Simon Schaffer, 207–233. Woodbridge: Boydell.

Oldroyd, David. 1996. *Thinking about the Earth: A History of Ideas in Geology.* London: Athlone Press; Cambridge, MA: Harvard UP.

Oldroyd, David. 2000. "James Hutton's "Theory of the Earth" (1788)." *E* 23: 196–202.

Oldroyd, David, and John Howes. 1978. "The First Published Version of Leibniz's *Protogœa*." *Journal for the Society of the Bibliography of Natural History* 9: 56–60.

Oliver, Jack. 1996. *Shocks and Rocks: Seismology in the Plate Tectonics Revolution.* Washington, DC: American Geophysical Union.

Opdyke, Neil, Billy Glass, James Hays, and John Foster. 1966. "Paleomagnetic Study of Antarctic Deep-Sea Cores." *S* 123: 349–357.

Oreskes, Naomi. 1999. *The Rejection of Continental Drift* . . . New York: Oxford UP.

Oreskes, Naomi, and Homer Le Grand, eds. 2001. *Plate Tectonics: An Insider's History* . . . Boulder, CO: Westview Press.

Ovidus, Publius. 1955. *The Metamorphoses of Ovid.* Harmondsworth: Penguin.

Payton, Charles, ed. 1977. *Seismic Stratigraphy—Applications to Hydrocarbon Exploration.* Tulsa: American Association of Petroleum Geologists.

Penck, Albrecht. 1882. *Die Vergletscherung die Deutscher Alpen* . . . Leipzig: Barth.

Penck, Albrecht. 1894. *Morphologie der Erdoberfläche.* Stuttgart: Engelhorn.

Penck, Albrecht, and Eduard Brückner. 1901–1909. *Die Alpen im Eiszeitalter.* 3 vols. Leipzig: Tauchnitz.

Penck, Walther. 1924. *Die morphologische Analyse: Ein Kapitel physikalischen Geologie.* Stuttgart: Geographische Abhandlungen.

Penck, Walther. 1925. "Die Piedmontflächen des südlichen Schwarzwaldes." *Zeitschrift der Gesellschaft für Erdkunde zu Berlin*, 83–108.

Penck, Walther. 1953. *Morphological Analysis of Land Forms . . .* London: Macmillan.

Pigott, John. 1981. "Global Tectonic Control of Secular Variations in Phanerozoic Sedimentary Rock/Ocean/Atmosphere Chemistry." Ph.D. diss., Northwestern University.

Plato. 1970. *The Dialogues of Plato* 1. London: Sphere Books.

Playfair, John. 1802. *Illustrations of the Huttonian Theory of the Earth.* London: Cadell and Davies; Edinburgh: Creech.

Playfair, John. 1805. "Biographical Account of the Late Dr James Hutton . . . ," *Transactions of the Royal Society of Edinburgh* 3: 39–99.

Powell, John. 1875. *Exploration of the Colorado River . . .* Washington, DC: Government Printing Office.

Prévost, Constant. 1839–1840. "Opinion sur la théorie des soulèvements." *Bulletin de la Société Géologique de France* 10: 430; 11: 183–203.

Ramsay, Andrew. 1862. "On the Glacial Origin of Certain Lakes in Switzerland . . . and Elsewhere." *QJGS* 18: 185–204 and plate.

Ramsbottom, William. 1977. "Major Cycles of Transgression and Regression (Mesothems) in the Namurian." *Proceedings of the Yorkshire Geological Society* 41: 261–291.

Rappaport, Rhoda. 1986. "Hooke on Earthquakes: Lectures, Strategy and Audience." *BJHS* 19: 129–146.

Raup, David, and John Sepkoski. 1984. "Periodicity of Extinctions in the Geological Past." *Proceedings of the National Academy of Sciences* 81: 801–805.

Read, Herbert. 1957. *The Granite Controversy.* London: Murby.

Reade, T. Mellard. 1877. "President's Address." *Proceedings of the Liverpool Geological Society* 3: 211–235

Rhode, Robert, and Richard Muller. 2005. "Cycles in Fossil Diversity." *N* 434: 208–210.

Robertson, Thomas. 1952. "Plant Control in Rhythmic Sedimentation." *Compte rendu du 3e Congrès de Stratigraphie et Géologie du Carbonifère, Heerlen*, 515–521.

Rogers, John. 1996. "A History of the Continents in the Past Three Billion Years." *JG* 104: 91–107.

Rudwick, Martin. 1969. "Lyell on Etna, and the Antiquity of the Earth." In *Toward a History of Geology*, ed. Cecil Schneer, 288–304. Cambridge, MA: MIT Press.

Rudwick, Martin. 1971. "Uniformity and Progression: Reflections on the Structure of Geological Theory in the Age of Lyell." In *Perspectives in the History of Science and Technology*, ed. Duane Roller, 209–227. Norman: University of Oklahoma Press.

Rudwick, Martin. 1977. "Historical Analogies in the Early Geological Work of Charles Lyell." *Janus* 64: 89–107.

Rudwick, Martin. 1978. "Charles Lyell's Dream of a Statistical Palaeontology." *Palaeontology* 21: 225–244.

Rudwick, Martin. 2005. *Bursting the Limits of Time: The Reconstruction of Geohistory in the Age of Revolution.* Chicago and London: Chicago UP.

Runcorn, Keith. 1962. "Palaeomagnetic Evidence for Continental Drift and Its Geophysical Cause." In *Continental Drift*, ed. Keith Runcorn, 1–40. New York: Academic Press.

Russell, R. J. 1958. "Geological Morphology." *BGSA* 69: 1–12.

Schidlowski, Manfred, and Christian Junge. 1981. "Coupling among the Terrestrial Sulfur, Carbon, and Oxygen Cycles: Numerical Modelling Based on Revised Phanerozoic Carbon Isotope Record." *Geochimica et cosmochimica acta* 45: 589–594.

Schönbein, Christian. 1840. "On the Causes of the Change of Colour Which Takes Place in Certain Substances under the Influence of Heat." *Annals of Electricity, Magnetism and Chemistry* 5: 224–236 (first published in *Annalen der Physik und Chemie*, 1838).

Schuchert, Charles. 1932. "Gondwana Land Bridges." *BGSA* 43: 875–916.

Schwarzacher, W. 1993. *Cyclostratigraphy and the Milankovitch Theory.* Amsterdam: Elsevier.

Schwarzbach, Martin. 1986. *Alfred Wegener: The Father of Continental Drift.* Madison, WI: Science Tech Publishers; Berlin: Springer.

Seneca, Lucius. 1971–1972. *Naturales quaestiones.* 2 vols. London: Heinemann; Cambridge, MA: Harvard UP.

Şengör, Celâl. 1982. "Eduard Suess' Relations to the Pre-1950 School of Thought in Global Tectonics." *Geologische Rundschau* 71: 381–420.

Şengör, Celâl. 2003. *The Large-Wavelength Deformations of the Lithosphere: Materials for a History of the Evolution of Thought from the Earliest Times to Plate Tectonics.* Boulder, CO: GSA.

Serbian Academy of Sciences and Arts. 2004. *Milutin Milankovitch Anniversary Symposium: Paleoclimate and the Earth Climate System Belgrade, Serbia, 30 August–2 September 2004 Abstracts.* Belgrade.

Shackleton, Nicholas, André Berger, and Richard Peltier. 1990. "An Alternative Astronomical Calibration of the Lower Pleistocene Time-Scale Based on O[cean] D[rilling] P[rogram] Site 677." *Transactions of the Royal Society of Edinburgh: Earth Sciences* 81: 251–261.

Shackleton, Nicholas, and Neil Opdyke. 1973. "Oxygen Isotope and Palaeomagnetic Stratigraphy of Equatorial Pacific Core V28–238: Oxygen Isotope Temperatures and Ice Volumes on a 10^5 and 10^6 Year Scale." *Quaternary Research* 3: 39–55.

Shapiro, Herman, ed. 1964. *Medieval Philosophy: Selected Readings from Augustine to Buridan.* New York: Random House.

Shepard, Francis. 1978. "Memorial to Phillip H. Kuenen." *Memorials of the GSA* 8: 1–2.

Sloss, Laurence. 1950. "Paleozoic Sedimentation in Montana Area." *BAAPG* 34: 423–451.

Sloss, Laurence. 1963. "Sequences in the Cratonic Interior of North America." *BGSA* 74: 93–114.

Sloss, Laurence. 1972. "Synchrony of Phaenrozoic Sedimentary–Tectonic Events of the North American Craton and the Russian Platform," *International Geological Congress . . . Canada–1972. Section 6 . . .* , 24–32. Montreal.

Sloss, Laurence. 1988. "Forty Years of Sequence Stratigraphy." *BGSA* 100: 93–114.

Sloss, Laurence. 1991. "The Tectonic Factor in Sea-Level Change: A Countervailing View." *Journal of Geophysical Research* 96: 6609–6617.

Sloss, Laurence, William Krumbein, and Edward Dapples. 1949. "Integrated Facies Analysis." GSA Memoir 39: 91–124.

Smith, Richard. 1703. *Bibliotheca Hookiana* ... London: Millington and Smith.

Steno, Nicolaus. 1968. *The Prodromus of Nicolaus Steno's Dissertation concerning a Solid within a Solid.* New York: Hafner.

Steno, Nicolaus. 1669. *De solido intra solidum naturaliter contento dissertatione Prodromus.* Florence: Signo Stellae.

Stille, Hans. 1924. *Grundfragen der vergleichenden Tektonik.* Berlin: Borntraeger.

Strabo. 1959. *The Geography of Strabo,* 8. London: Heinemann; Cambridge, MA: Harvard UP.

Strabo. 1960. *The Geography of Strabo,* 1. London: Heinemann; Cambridge, MA: Harvard UP.

Suess, Eduard. 1875. *Die Entstehung der Alpen.* Vienna: Braumüller.

Suess, Eduard. 1885–1909. *Das Antlitz der Erde.* 3 vols. Prague: Tempsky and Freytag.

Suess, Eduard. 1897–1918. *La face de la terre.* 3 vols. Paris: Librairie Armand Colin.

Suess, Eduard. 1904–1924. *The Face of the Earth.* 5 vols. Oxford: Clarendon Press.

Suzuki, Yasumoto. 2001. "Kiyoo Wadati and the Path to the Discovery of the Intermediate–Deep Earthquake Zone." *E* 24: 118–123.

Tasch, Paul. 1975. "A Quantitative Estimate of Geological Time by Lyell." *Isis* 66: 406.

Tasch, Paul. 1977. "Lyell's Geochronological Model: Published Year Values for Geological Time." *Isis* 68: 440–442.

Tomkeieff, Sergei. 1946. "James Hutton's 'Theory of the Earth,' 1795." *Proceedings of the Geologists' Association* 57: 322–328.

Tomkeieff, Sergei. 1948. "James Hutton and the Philosophy of Geology." *Transactions of the Edinburgh Geological Society* 14: 253–276.

Tomkeieff, Sergei. 1962. "Unconformity—An Historical Study." *Proceedings of the Geologists' Association* 73: 383–417.

Turner, Anthony. 1974. "Hooke's Theory of the Earth's Axial Displacement: Some Contemporary Opinions." *BJHS* 7: 166–170.

Udden, Johann. 1912. "Geology and Mineral Resources of the Peoria Quadrangle, Illinois." *USGSB* 506. Washington, DC: Government Printing Office.

Ulrich, Edward. 1911. "Revision of the Paleozoic Systems." *BGSA* 22: 281–680.

Umbgrove, Johannes. 1947. *The Pulse of the Earth.* The Hague: Nijhoff.

Urey, Harold. 1947. "The Thermodynamic Properties of Isotopic Substances." *Journal of the Chemical Society,* 562–581.

Usher, James. 1658. *Annals of the World . . .* London: Crook (1st Latin ed., 1650).

Vail, Peter. 1992. "The Evolution of Seismic Stratigraphy and the Global Sea-Level Curve." In *Eustasy . . .* ed. Robert Dott, 83–91. GSA Memoir 180. Boulder, CO: GSA.

Vail, Peter, F. Audemard, S. A. Bowman, P. N. Eisner, and C. Perez-Cruz. 1991. "The Stratigraphic Signatures of Tectonics, Eustasy and Sedimentology An Overview." In *Cycles and Events in Stratigraphy,* ed. Gerhard Einsele, Werner Ricken, and Adolf Seilacher, 617–659. Berlin: Springer.

Vail, Peter, Robert Mitchum, and S. Thompson III. 1977a. "Seismic Stratigraphy and Global Changes of Sea Level, Part 3: Relative Changes of Sea Level from Coastal Onlap." In *Seismic Stratigraphy . . .* ed. Charles Payton, 63–81. Tulsa, OK: AAPG.

Vail, Peter, Robert Mitchum, and S. Thompson III. 1977b. "Seismic Stratigraphy and Global Changes of Sea Level, Part 4: Global Cycles of Relative Changes of Sea Level." In *Seismic Stratigraphy . . .* ed. Charles Payton, pp. 83–97. Tulsa, OK: AAPG.

Vail, Peter, Robert Mitchum, Robert Todd, J. Michael Widmier, S. Thompson III, John Sangree, J. Bubb, and W. Hatlelid. 1977. "Seismic Stratigraphy and Global Changes of Sea Level." In *Seismic Stratigraphy . . .* ed. Charles Payton, pp. 47–212. Tulsa, OK: AAPG. (This publication appeared as essentially 11 separate papers.)

Vail, Peter, and Robert Wilbur. 1966. "Onlap, Key to Unconformities and Depositional Cycles." *BAAPG* 50: 638–639.

Varen[ius], Bernhard. 1650. *Geographia generalis.* Amsterdam: Elzeverium.

Vernadsky, Vladimir. 1924. *La géochimie.* Paris: Alcan.

Vernadsky, Vladimir. 1929. *La biosphère.* Paris: Alcan.

Vernadsky, Vladimir. 1997. *The Biosphere.* New York: Springer-Verlag.

Vine, Frederick, and Drummond Matthews. 1963. "Magnetic Anomalies over Oceanic Ridges." *N* 199: 947–949.

Wadati, Kiyoo. 1935. "On the Activity of Deep-Focus Earthquakes in the Japan Islands and Neighbourhoods." *Geophysical Magazine* 8: 305–325.

Walker, Gabriel. 2003. *Snowball Earth.* London: Bloomsbury Publishing.

Waller, Richard, ed. 1705. *The Posthumous Works of Robert Hooke.* London: Smith and Walford.

Walther, Johannes. 1893–1894. *Einleitung in die Geologie als historische Wissenschaft.* Jena: Fischer. Vol. 1: *Bionomie des Meeres . . . ,* 1–196; vol. 2: *Die Lebensweise der Meeresthiere . . . ,* 200–531; vol. 3: *Lithogenesis der Gegenwart . . . ,* 535–1055.

Wanless, Harold. 1931. "Pennsylvanian Cycles in Western Illinois." *Illinois State Geological Survey,* Bulletin 60: 179–193.

Wanless, Harold, and Julie Cannon. 1966. "Late Paleozoic Glaciation." *Earth-Science Reviews* 1: 247–286.

Wanless, Harold, and Francis Shepard. 1936. "Sea Level Changes and Climatic Changes Related to Late Paleozoic Cycles." *BGSA* 97: 1177–1206.

Wanless, Harold, and James Weller. 1932. "Correlation and Extent of Pennsylvanian Cyclothems." *BGSA* 43: 1003–1016.

Wegener, Alfred. 1912. "Die Entstehung de Kontinente." *Petermann's Mitteilungen aus Justus Perthes' Geographischer Anstalt* 58: 185–195, 253–256, 305–309.

Wegener, Alfred. 1915. *Die Entstehung der Kontinente und Ozeane.* Braunschweig: Vieweg.

Wegener, Alfred. 1966. *The Origin of Continents and Oceans.* London: Methuen.

Weller, Marvin. 1930. "Cyclical Sedimentation of the Pennsylvanian Period and Its Significance." *JG* 38: 97–135.

Weller, Marvin. 1931. "The Conception of Cyclical Sedimentation during the Pennsylvanian Period." *Bulletin of the Illinois State Geological Survey* 36: 163–177.

Weller, Marvin. 1942. "Rhythms in Upper Pennsylvanian Cyclothems." *Transactions of the Illinois Academy of Science* 75: 145–146.

Weller, Marvin. 1956. "Argument for Diastrophic Control of Late Paleozoic Cyclothems." *BAAPG* 40: 17–50.

Weller, Marvin. 1958. "Cyclothems and Larger Sedimentary Cycles of the Pennsylvanian." *JG* 66: 195–207.

Wells, Alan. 1960. "Cyclic Sedimentation: A Review." *Geological Magazine* 97: 389–403.

Werner, Abraham. 1971. *Short Classification and Description of the Various Rocks,* trans. Alexander M. Ospovat. New York: Hafner.

Westbroek, Peter. 1992. *Life as a Geological Force ...* New York: Norton.

Wheeler, Harry. 1958. "Time-Stratigraphy". *BAAPG* 42: 1047–1063.

Wheeler, Harry. 1959. "Stratigraphic Units in Space and Time." *AJS* 257: 692–706.

Wiechert, Emil. 1897. "Über die Massenverteilung im Innern der Erde." *Nachrichten von der Königliche Gesellschaft der Wissenschaften zu Göttingen. Mathematisch-physikalischen Klasse,* 221–243.

Willman, Bowen, and Norman Payne. 1942. "Geology and Mineral Resources of the Marseilles, Ottawa, and Streator Quadrangles." *Bulletin of the State Geological Survey of Illinois* 66.

Wilson, Tuzo. 1965. "A New Class of Faults and their Bearing on Continental Drift." *N* 207: 343–347.

Wilson, Tuzo. 1966. "Did the Atlantic Close and Then Re-open?" *N* 211: 676–681.

Yamada, Toshihiro. 2003. "Stenonian Revolution or Leibnizian Revival? Constructing Geohistory in the Seventeenth Century." *Historia scientiarum* 13: 75–100.

Zeller, Edward. 1964. "Cycles and Psychology." In *Symposium on Cyclic Sedimentation,* ed. Daniel Merriam, 2 vols. Geological Survey of Kansas, Bulletin 169. 2: 631–636.

INDEX

About the Author

DAVID OLDROYD is Honorary Visiting Professor in Science and Technology Studies at the University of New South Wales. He was Secretary General of the International Commission on the History of Geological Sciences. He is the author of dozens of articles and several books, including *The Highlands Controversy: Constructing Geological Knowledge through Fieldwork in Nineteenth-Century Britain* and *Thinking About the Earth: A History of Ideas in Geology*.